NAVY SEALS BUG IN GUIDE

Transform Your Home into a Fortress of Absolute Security

Mark Reynolds

INDEX

INTRODUCTION

The value of bugging in: Why staying home is the safest course of action

When it comes to emergency or crisis situations, the temptation to flee immediately can be strong, especially for those with an adventurous spirit or a military background. However, one of the key concepts that Navy Seals learn through years of training and real missions is the importance of assessing the situation coolly and intelligently. In many circumstances, the safest and most strategic choice is not to flee, but to remain in a safe and well-defensible position, commonly known as "bugging in". Staying home during a crisis is an often overlooked course of action, but it's actually the one that offers the most control and security. Your home is your territory. You know every corner, every access point, and you already have everything you need at your fingertips.

Unlike an impromptu evacuation, which takes you into an uncertain and potentially hostile environment, staying at home allows you to use all the resources you have accumulated over time. You can maintain control over food, water, and medical supplies, which can be difficult, if not impossible, to secure in a flight context. Another key aspect of bugging in is the ability to maintain a solid defensive line. At home you can organize defenses much more effectively than on the move, using strategic points and knowing escape routes and hiding places by heart. If you are well prepared, your home can transform into a fortress that protects you from external threats, be they intruders or adverse weather conditions.

Navy Seals learn to assess terrain and adapt quickly, but they also know that impulsive action can lead to unnecessary exposure to danger. In many cases, familiar ground is the safest. Furthermore, the psychological aspect of bugging in should not be underestimated. Staying in a familiar environment reduces stress and increases the ability to make rational decisions. Panic is one of the worst enemies in a crisis situation, and being in an unfamiliar place or in constant motion fuels this state of mind. At home, however, you can count on an environment you are familiar with and a certain level of comfort, factors that allow you to maintain clarity and calm, essential elements for successfully overcoming a crisis. Finally, staying at home gives you the ability to communicate more easily with those outside, both to receive updated information and to coordinate with friends, family or local communities. In a bug out environment, communications can be difficult, and fragmented information can lead to poor decisions.

Lessons from a Navy Seal: Practical Survival Strategies

When you think of a Navy Seal, the image of a highly trained soldier immediately comes to mind, capable of facing the most extreme and dangerous situations with coolness and determination. However, what is often overlooked is that many of the survival strategies these elite men employ are not just reserved for the battlefield. They are practical lessons that can be applied in everyday life to deal with crises and difficult situations with the same effectiveness. The first fundamental lesson a Navy Seal learns is mental preparation. Survival is not just a question of physical strength, but above all of mental resilience. Seals are trained to remain calm even in the most intense situations, because they know that panic is their worst enemy. In any emergency scenario, the ability to maintain a clear head is what makes the difference between life and death. In daily life, this translates into the need to develop a resilient mentality, capable of facing unexpected events without being overwhelmed by anxiety or fear. Another key Navy Seal strategy is the importance of advance planning and preparation.

Before any mission, every detail is analyzed and predicted, and contingency plans are prepared for every possible scenario. This principle can also be applied in civilian life: being prepared for unexpected events, such as a natural disaster or a family crisis, can make a huge difference. Preparing an emergency kit, having an evacuation plan, and knowing the resources available in your area are all measures that can significantly improve your chances of survival in a crisis situation. The ability to adapt is another vital lesson learned by Navy Seals. During operations, conditions can change in an instant, and flexibility is essential to survival. Being able to quickly adapt to new information and modify plans accordingly is one of the most valuable skills a Seal possesses. In everyday life, this translates into the ability to not get stuck in a single way of thinking. Knowing how to recognize when a situation requires a change in approach and having the courage to implement it can be life-saving, both literally and figuratively. The importance of teamwork is another lesson that Navy Seals learn thoroughly.

Despite the image of lone warriors often associated with them, the Seals know that success depends on the ability to work together as a cohesive unit. In any survival situation, being able to rely on others, share resources and expertise, and support each other can greatly increase your chances of overcoming the crisis. In civilian life, this means cultivating strong relationships with family, friends and community, creating support networks that can be crucial in difficult times. Navy Seals teach the importance of physical resilience. Being in good physical shape is not just a question of vanity, but a real necessity for survival. The ability to resist fatigue, pain and physical stress is essential in emergency situations. Maintaining a fit and healthy body prepares you not only to face daily challenges, but also to face unexpected events that require strength and endurance. The survival lessons learned by Navy Seals go far beyond the battlefield. These are practical strategies that can be applied by anyone to face life's adversities with more confidence and determination. Mental preparation, planning, adaptability, teamwork and physical resilience are the keys to overcoming difficulties, whether it is a life crisis or an everyday situation that requires alertness and strength. These lessons remind us that survival is not just about overcoming extreme difficulties, but also about living more consciously and preparedly every day.

CHAPTER 1

Prepare your home for any emergency

Preparing your home for any emergency is an act of responsibility that requires attention to detail and a mindset ready to face the unexpected. Every home, regardless of its location or size, should be equipped to withstand emergency situations that could range from natural disasters, such as earthquakes or floods, to more prolonged crises, such as power outages or supply problems. The first thing to consider is safety. It is essential to know the critical points of your home, such as escape routes, safe places in case of collapses and first aid tools. Each family member must be aware of these aspects and ready to act calmly and decisively.

It is important to have sufficient supplies of water and non-perishable food available, which can sustain the family for at least two weeks. These supplies should be stored in a location that is easily accessible but protected from potential damage. A well-stocked emergency kit is essential: it should include medicines, flashlights with spare batteries, a hand-crank or battery-powered radio to receive updates, thermal blankets and multi-purpose tools. In the event of an emergency, communication could become an issue, so it is helpful to have a contact plan with family and friends, as well as alternative communication devices such as two-way radios.

Energy is another crucial aspect. Considering installing a portable generator or solar panels can make a difference in extended blackout situations. The protection of the house itself should not be neglected: reinforcing doors and windows, making sure the roof is solid and periodically checking the integrity of the structure are preventive actions that can minimize damage. The right mentality is to foresee the unexpected and prepare to respond with determination. The home must not become a passive refuge, but a secure operational base from which to face every challenge. Being prepared does not mean living in fear, but having the awareness and ability to protect yourself and your loved ones in any circumstance.

Recognize possible dangers

Recognizing the possible dangers inside your home is the first step to transforming it into a safe and protected place. Every environment, however familiar it may seem, hides pitfalls which, if not addressed in time, can turn into real threats. Home security is not limited to protecting yourself from intruders or catastrophic events, but includes a 360-degree vision that encompasses every aspect of daily life. The first thing to do is to develop active awareness, a state of vigilance that leaves no room for superficiality. Every object, every access, every vulnerable point must be evaluated with a critical eye. Doors and windows are obviously the first elements to consider: they must be solid, well closed and, if possible, reinforced. But that's not enough, we also need to think about secondary accesses, such as garages, cellars or basement windows, often overlooked but equally crucial.

The home's electrical, plumbing and heating systems represent other potential sources of danger. A short circuit, a water leak or a heating malfunction can trigger fires, floods or poisoning. It is essential to keep these systems in perfect condition, carrying out regular checks and not putting off necessary repairs. The layout of the internal spaces must also be designed to minimize risks: avoid clutter that could obstruct escape routes, pay attention to pointed or sharp objects and make sure that chemical or toxic products are stored in safe places, out of reach. of children.

Another aspect to consider is protection from external events. If the house is located in an earthquake-prone area, for example, it is advisable to anchor heavy furniture and install safety systems to prevent collapses. However, if the danger comes from outside, such as in the case of storms or floods, it is essential to have adequate physical barriers or drainage systems. Being aware of the threats that may arise is only half the battle. The next step is to act with determination to neutralize them, eliminating every weak point and transforming the house into a safe haven, where every detail has been thought of to guarantee the protection and tranquility of those who live there.

Being prepared means anticipating the enemy's moves, which in this case can be an intruder, a natural event or simple carelessness. It is a game of prevention, in which nothing can be left to chance. The safety of your home starts with your ability to see beyond appearances, spot what can go wrong and intervene before it becomes a problem. It is a constant operation, which requires discipline, but it is also the only way to guarantee you and your loved ones the peace and security they deserve.

Conducting a home security assessment

Carrying out a home security assessment is a crucial step in ensuring that your home is a safe place, capable of protecting you and your loved ones from any threat. This process requires meticulous attention to detail, a strategic approach, and the ability to identify and address potential vulnerabilities. It's not just a matter of checking that doors and windows are closed properly, but of going further, examining every aspect of the house with the critical eye of someone who knows that security is not a luxury, but a necessity.

The assessment starts from the outside, looking at the property as someone trying to enter would. You need to consider every possible access point, every hidden corner, every bush that could offer cover to an intruder. The perimeter of your home should be your first bastion of defense, and that means keeping the grounds clean and well-lit, installing security cameras if necessary, and making sure fences are solid and free of weak points. But it's not enough to stop here. Security also extends inside the home, where every room must be evaluated with the same attention.

Identify areas that could pose a risk, such as stairs without handrails, exposed electrical outlets or carpets that can cause slips. It is also important to evaluate the quality of locks, alarm systems and emergency lights, which must always be functional and easily accessible. Each room should have an emergency plan, a clear and obstacle-free escape route. The kitchen, often the heart of the home, is also one of the most dangerous areas. Here, a safety assessment must consider the location of knives, the stability of appliances, and the presence of working smoke and carbon monoxide detectors.

Another crucial element of the assessment is the protection of data and personal information. In an age where cyber security has become just as important as physical security, it is crucial to ensure that your computers and mobile devices are protected by strong, up-to-date passwords, and that your Wi-Fi network is secure and not accessible to outsiders. Every aspect of home life must be considered, because security is a set of small details that, put together, create an effective and reliable protection system.

A home security assessment is an ongoing process, not a do-it-once-and-forget-it operation. Conditions change, as do potential threats. You need to regularly review your home, update safety measures and adapt to new situations. Only with this mentality, with this discipline, is it possible to ensure that your home remains a safe place, where you and your family can live in peace, knowing that you have done everything possible to protect yourself from any risk.

Formulate a customized Bug-In strategy

Formulating a personalized Bug-In strategy is essential to ensuring that you and your family are prepared to deal with any emergency situation within the walls of your home. The Bug-In concept involves staying in your home during a crisis, rather than evacuating, and requires detailed planning to ensure every aspect of survival and safety has been considered. The key to developing an effective strategy is to understand that there is no one-size-fits-all solution; every home, every family and every scenario requires a tailored approach, based on a careful assessment of the available resources, possible threats and the specific needs of those living under that roof.

The first step to formulating a customized Bug-In strategy is to analyze the context in which you find yourself. You need to have a thorough understanding of your environment: your neighborhood, local resources, natural and human hazards that may influence your decision to stay indoors. It's not just about thinking about catastrophic events like hurricanes or earthquakes, but also about more common and equally dangerous situations, like prolonged power outages, health crises or civil unrest. Once you understand the context, you need to consider the resources you have and those you need to procure. Water, food, energy and medicines are the foundation on which to build your plan, but it is equally important to have access to reliable information and functioning communication systems.

Your home must become a safe, self-sufficient haven, and to do so you must think about how to protect every resource, how to manage supplies, and how to keep your spirits high during periods of prolonged isolation. Each family member has a role to play in the Bug-In strategy, and the key to success lies in preparation and training. Everyone needs to know what to do in an emergency, where to find supplies, how to correctly use the tools available and, above all, how to stay calm under pressure. Communication is key, and must be clear and constant to avoid confusion and panic.

The physical security of the home is another critical element. You must consider how to protect the building from potential intruders, how to strengthen the main entrances and how to create physical or psychological barriers that will discourage any intrusion attempts. But it's not just about defending yourself from others: you also need to think about how to protect your family from themselves, making sure everyone knows how to safely handle any weapons or defense tools and how to avoid domestic accidents under stressful conditions.

A custom Bug-In strategy is never static; must be reviewed and updated regularly to account for changes in your personal situation, home setup, or threat landscape. It's a form of insurance against uncertainty, a plan that allows you to face the unknown knowing that you've done everything you can to protect yourself and your loved ones. Ultimately, the strength of a Bug-In strategy lies in your ability to predict, prepare and adapt, thus ensuring that your home remains a safe haven under all circumstances.

CHAPTER 2

Preserving food for the future

Food preservation is a topic of great importance, especially in an era where uncertainty and climate change can affect the availability and quality of food. Preparing for the future means thinking about how we can preserve food resources effectively, ensuring they are available and safe for an extended period. This practice is not only a safety measure, but also a way to reduce food waste, contributing to more sustainable management of the planet's resources. Food preservation has ancient roots, dating back to when humanity began to understand the importance of keeping food in good condition during times of scarcity. The techniques used today are the result of centuries of experimentation and innovation, but the principle remains the same: prolong the life of foods by protecting them from the factors that cause deterioration. This includes protection from microorganisms, oxidation and changes in temperature and humidity. One of the most common preservation techniques is refrigeration, which slows down the deterioration of food by lowering the temperature. Cold slows the growth of bacteria and other microorganisms that can cause decay.

However, refrigeration has its limitations: not all food can be stored for long in the refrigerator, and in the event of prolonged power outages, food can spoil quickly. Another preservation technique is drying, which removes moisture from foods, making it difficult for bacteria and mold to grow. This method has been used for millennia and is still one of the most effective techniques today, especially for preserving fruit, vegetables, meat and fish. Drying can occur naturally, with exposure to the sun, or through the use of electric dryers, which offer greater control over the process. Vacuum preservation is another increasingly widespread practice, which consists of eliminating air from food packages. This method reduces oxidation and slows the growth of microorganisms, significantly extending the life of foods. It is particularly useful for preserving meat, fish and cheese, but also for fruit and vegetables that you want to keep fresh for longer. Canning and preserving in jars are techniques that allow you to preserve food for years. This method involves cooking food at high temperatures to kill any microorganisms, followed by an airtight seal that prevents future contamination. This technique is particularly suitable for fruits, vegetables, sauces and soups, and is a favorite method for building long-term food reserves. Fermentation is another preservation method that not only preserves food but also enriches it with flavor and beneficial nutrients.

During fermentation, natural microorganisms transform sugars and starches into acids or alcohol, which act as natural preservatives. Foods like yogurt, sauerkraut, kimchi, and miso are examples of fermented foods that can last a long time and offer health benefits. Thinking about food preservation for the future also means thinking about how we can reduce food waste. Every year, tons of food are thrown away because they were not consumed in time. Learning to store food correctly not only allows us to have supplies for times of need, but also to better use the resources at our disposal, reducing the environmental impact. In conclusion, food preservation is a fundamental practice to ensure food safety and promote sustainability. The different techniques available offer multiple ways to preserve food, allowing us to face the future with greater tranquility and awareness. Preparing today means ensuring that, whatever happens, we will always have access to safe and healthy food for us and future generations.

Foods essential for survival

In emergency situations, having access to essential foods can mean the difference between maintaining strength and health, or succumbing to hardship. Foods essential for survival are not only those that provide nutrition, but also those that can be stored for a long time, easily transported and prepared simply. The choice of which foods to have available must be well thought out, taking into account nutritional needs, shelf life and availability in critical situations. The first criterion to consider when selecting foods essential for survival is nutritional value. In an emergency situation, it is essential to have available foods that provide the energy, proteins, vitamins and minerals needed to sustain the body. Foods rich in complex carbohydrates, such as rice, pasta and cereals, are important because they provide long-term energy, essential for dealing with prolonged physical and mental efforts. At the same time, protein, found in foods like beans, legumes, nuts and jerky, is critical for maintaining muscle mass and supporting the immune system. In addition to nutritional value, another key factor is shelf life. Survival foods must be able to be stored for long periods without spoiling. This is especially important in scenarios where access to fresh food may be limited or non-existent.

Foods such as rice, dried pasta, cans of vegetables, fruit and meat, as well as freeze-dried products, are ideal because they can last for months or even years if stored correctly. These foods not only guarantee a continuous supply of nutrients, but also offer the security of having sufficient supplies to face prolonged periods of difficulty. Convenience and ease of preparation are other essential considerations when choosing foods for survival. In emergency situations, access to cooking utensils, electricity or gas may be limited. Therefore, it is important to have foods that require little or no preparation, such as canned convenience foods, energy bars or dried fruit. These foods can be eaten directly without the need for cooking, which is especially useful when resources are scarce.

Even water, necessary for the preparation of many foods, can be a limited good; therefore, foods that require little water to prepare are preferable. Another aspect to consider is variety. While the priority is to have nutritious, long-lasting foods, it is important to include some variety in your survival diet to avoid nutritional deficiencies and to keep your spirits high. Eating the same food for days or weeks can become monotonous and demotivating. Including a range of flavors and textures, such as nuts, chocolate, seasonings and spices, can help make a difficult period more bearable. This not only helps maintain appetite, but also contributes to psychological well-being, which is crucial in stressful situations. In conclusion, foods essential for survival must be chosen carefully, taking into account nutritional value, shelf life, ease of preparation and variety. Having an adequate supply of these foods can make the difference between overcoming a crisis situation safely and with dignity, or facing unnecessary hardship.

Preparing in advance, carefully selecting the foods to include in your emergency supplies, is a fundamental step in ensuring survival in an emergency, offering both physical and psychological support during difficult times.

Creation and replacement of food supplies

Food supplies are a crucial component of this approach to home security, allowing you to be self-sufficient in times of crisis, when access to food may be limited or impossible. Building food supplies does not simply mean accumulating food in large quantities. It's about planning intelligently, selecting foods that not only can last over time, but which are also suitable for providing the right nutritional intake. Each food you choose must be part of a larger strategy, which includes the possibility of addressing different situations: from temporary interruptions in the food supply, to more complex scenarios where you may have to rely on these reserves for an extended period. Once you've built up an adequate supply, the key to keeping your home truly safe is ongoing management of these resources. It's not enough to fill a pantry and forget about it. Foods have a limited shelf life, even those designed to last a long time. It is essential to develop a stock rotation system, which allows you to use older foods before they expire, replacing them with new products. This way, you not only avoid waste, but you ensure that your reserves are always fresh and ready to use.

The inventory replacement process should be an integral part of your routine. Establishing a calendar to regularly check expiration dates and update your reserves is essential. This approach allows you to always have food available that you can use immediately, without the risk of having to face a crisis with expired or inadequate food. Whenever you replace a product, be sure to note the purchase date and expiration date, so you can easily track when it will need to be used or replaced. Don't forget the importance of diversifying your stocks. Including a variety of foods not only improves the overall nutritional value of your supply, but also allows you to maintain a certain level of comfort and familiarity during times of emergency. Variety will help prevent food boredom and ensure you and your family can maintain a balanced diet, even when outside resources are scarce. Consider that preparation isn't just about food. Having an adequate supply of drinking water and other basic necessities is equally vital. Make sure you have a comprehensive plan that covers all your basic needs, because a truly secure home is one that can support you and your loved ones in any circumstance, without having to depend on external resources. Making your home a safe place means thinking ahead and acting decisively. Creating and managing food supplies is part of a broader framework of preparedness that will give you the confidence to face any situation with confidence and security. Don't leave anything to chance. Prepare your supplies carefully and manage them carefully, and your home will become the safe haven you need, at all times.

Advice on appropriate shelf life and storage

Understanding the shelf life and how to properly store food is essential to maintaining a safe and well-prepared environment. When we talk about food preservation, it's not just about keeping foods fresh, but about preserving their quality and nutritional value over time, so that they are ready to be used when needed. The shelf life of a food is influenced by several factors, including the type of product, the storage conditions and the type of packaging used. Understanding these aspects allows you to better organize your food supplies, ensuring that foods are still edible and nutritious when you need them. Some foods, such as rice, pasta and dried legumes, have a very long shelf life when stored in appropriate conditions, while others, such as fresh produce or meats, require more careful handling to prevent them from spoiling. Ideal food storage conditions vary, but in general, a cool, dry, dark environment best preserves the quality of most products. Temperature plays a crucial role; keeping food at a constant temperature, preferably below 20°C, helps prevent the growth of bacteria and mold. Humidity is another enemy of conservation, as it can accelerate decomposition and encourage the proliferation of microorganisms. Using airtight containers and materials that protect from light and moisture, such as dark glass or sturdy plastic, can make a big difference in how long your supplies last. Another important consideration is the storage method.

Refrigeration and freezing are effective techniques for extending the shelf life of many foods, but they depend on the constant availability of electricity, which may not be guaranteed in all situations. It is therefore essential to have an alternative plan, which includes foods that can be stored at room temperature without compromising their safety. In this context, sous vide, canning and freeze-drying are valuable methods for keeping foods edible and safe for long periods. An often overlooked aspect is inventory rotation. Even long-life products have an expiry date, and to ensure that your reserves are always usable, you need a system that allows you to consume older foods and replace them with new ones. This practice not only reduces waste, but ensures that you always have quality food, ready to use at any time. Taking note of expiration dates and keeping an updated inventory of supplies will help you stay on top of the situation, avoiding unwanted surprises at critical moments. It is important to remember that the safety of your home also depends on your ability to adapt to circumstances. Having well-maintained inventories is only part of the overall strategy. You must be ready to handle any emergencies without depending on external factors, and this requires careful and meticulous preparation. Knowing food shelf life and proper preservation techniques will give you peace of mind knowing that no matter what happens, you'll be ready to deal with it. Your home, well equipped and organized, will become your safe haven, capable of supporting and protecting you even in the most difficult moments.

CHAPTER 3:

Ensure access to water security

When we talk about home security, we often tend to think of alarm systems, sturdy locks and outdoor lighting. However, one crucial aspect that is often overlooked is water safety. Ensuring your home has access to a clean, safe water source is critical to protecting your family's health and well-being. Water is life, and without a reliable and safe supply, every other aspect of home security becomes meaningless.

The first thing to consider is the quality of the water that comes into your home's taps. Even if you live in an area served by a public water system, it's important to remember that the water may pick up contaminants along its path. Installing a filtration system is an essential step in eliminating impurities such as chlorine, lead, and other heavy metals that may be present in the water. Activated carbon or reverse osmosis filters are great options, each with their own specific benefits. While the activated carbon filter is effective at removing organic contaminants and chlorine, reverse osmosis offers broader protection, eliminating even the smallest contaminants.

Another critical aspect of water security is the protection of water supplies during emergencies or natural disasters. In situations where the water supply may be interrupted, having a supply of drinking water is vital. Make sure you always have at least three days' worth of water available for each family member, calculating around four liters per person per day. This supply must be stored in safe containers, away from sources of heat and direct light, and must be checked periodically to ensure its freshness. Investing in emergency water tanks or portable purification solutions can make all the difference in critical situations.

Another essential measure involves preventing flooding and water damage inside your home. Installing water leak detectors and automatic shutoff systems can prevent extensive damage to your home's structures and water quality. These systems can detect minor leaks that may go unnoticed, but over time could compromise the structural safety of your home and the quality of your living environment.

Finally, don't underestimate the importance of educating all family members about safe water practices. Teaching children and adults the importance of not wasting water, of immediately reporting any anomalies or leaks, and of following the correct procedures for conserving and purifying water is essential to ensure that everyone is prepared to handle any emergency situations. emergency.

Ensuring access to safe water in your home is not just a matter of prevention, but a fundamental duty to protect the health and lives of those who live there. Every gesture, every precaution taken to ensure that the water entering your home is clean, safe and always available, strengthens the very foundations of home safety. There is no room for error when it comes to your family's health, and water safety must be an uncompromising priority.

Assessment of your home's water needs

If we have to refer to home security, then one of the most crucial aspects is water management. Nothing can be left to chance when we talk about vital resources. Assessing your home's water needs requires meticulous planning and a precise understanding of what is needed to ensure the safety and well-being of those who live there. First of all, daily consumption must be taken into consideration. Each person in the house needs a specific amount of water for daily needs, ranging from drinking and personal hygiene to preparing meals and cleaning. Knowing your average requirement allows you to accurately calculate how much water is needed in case of emergencies or supply interruptions. In critical situations, such as a natural disaster or infrastructure failure, having an adequate supply of water can make the difference between being prepared and finding yourself in difficulty. It's not just about storing water, but doing it the right way, taking into account the duration of storage and storage conditions.

In addition to your daily needs, it is essential to consider the particular needs of your home. If you have a garden, an irrigation system, or if you use water for heating, all of these factors will influence the total amount you will need. Every home is different, and an accurate assessment of your water needs must be customized to your specific situation. This is where the importance of regularly monitoring your consumption and adjusting your reserves accordingly comes into play. You can't afford to overlook the details when safety is at stake. You also need to think about the infrastructure of your home. Efficient pipe systems, well-maintained water tanks and proper maintenance are key to ensuring that the water you use is always available and safe. Any neglect in this area can lead to dangerous situations, such as water contamination or leaks that put the entire facility at risk.

Assessing your home's water needs means preparing for every eventuality, with a mentality that leaves no room for improvisation. You need to plan ahead, know exactly how much water you need and make sure your supplies are adequate and easily accessible. This approach allows you to face any emergency with the security that comes from knowing you are ready. Safety is never a coincidence, it is the result of precise planning and constant preparation. When we talk about water, planning is everything, and there is no room for error. Your home should be a safe haven, and that means making sure your water needs are always under control, ready to respond to any challenge.

Solutions for safe water conservation

No matter how prepared we think we are, without a solid plan to ensure access to a safe water supply, any emergency can quickly turn into a crisis situation. Water is an essential resource, not only for daily survival, but also for maintaining hygiene and health in difficult situations. It is essential to have a storage system that not only meets basic needs, but is also robust, reliable and able to withstand unexpected events.

Choosing water containers is the first step. It is necessary to opt for high quality materials, which are safe for food contact and resistant over time. Food-grade polyethylene is one of the best options available: durable, lightweight, and doesn't leach harmful chemicals into the water. In addition to the material, it is important to consider the capacity of the containers. It's not just about having a sufficient quantity to cover immediate needs, but also about thinking about the long term, ensuring a supply that can last days, if not weeks, in the event of an emergency.

The placement of containers is equally crucial. They should be stored in a cool, dry place and away from direct sunlight. This reduces the risk of bacterial growth and extends the life of stored water. However, it's not enough to just stash water in a corner and forget about it - regularly rotating your supplies is a key aspect of safe storage. This means checking containers periodically, making sure they are well sealed and replacing the water with appropriate frequency, to ensure it is always fresh and drinkable.

In emergency situations, accessibility to water is essential. There is no point in having large supplies if they are not easily accessible. The containers must be positioned so that, if necessary, they can be transported and used quickly. Having an emergency plan that provides safe and accessible routes to water supplies is just as important as conserving them.

Finally, it is essential to consider alternative solutions for water purification, in case the initial supplies run out. Portable filters, disinfectant tablets or gravity systems are essential tools to have on hand. These devices allow you to transform potentially dangerous water sources into safe resources, protecting your family's health even in the most extreme situations.

Safely conserving water is a matter of discipline and planning. Nothing can be left to chance, every detail must be carefully considered, because when water is scarce, every mistake can cost you dearly. Your home must be a bastion of security, and to do this, water management must be impeccable. There is no room for superficiality. Safety requires constant commitment, vigilance and preparation that knows no compromises.

Filtration and purification methods

A fundamental aspect, which can make the difference in emergency situations, is the ability to guarantee access to safe and drinkable water. It doesn't matter how sturdy your walls are or how advanced the technology protecting your home is: if you don't have clean water, your safety is at risk. In scenarios where water resources are compromised, having effective filtration and purification methods can represent the dividing line between well-being and danger.

The first step is to understand that not all water is the same. Even what appears clear may contain dangerous contaminants, such as bacteria, viruses, heavy metals or harmful chemicals. This is where filtration systems come into play, representing the first line of defense. There are different technologies, each with its own specific field of action. Activated carbon filters, for example, are extremely effective at absorbing organic impurities and chlorine, improving the taste and odor of water. However, they don't remove everything. To deal with smaller contaminants like viruses and heavy metals, you need to turn to more sophisticated methods, such as reverse osmosis filtration. This process uses a semi-permeable membrane to retain even the smallest particles, ensuring almost pure water.

However, filtration alone is not always sufficient. In emergency situations, when water may come from unsafe sources such as rivers, lakes or even puddles, it is crucial to combine filtration with purification. Purification methods, such as boiling, disinfectant tablets or UV rays, go beyond simply removing physical impurities: they destroy pathogenic microorganisms that may be present in the water. Boiling is one of the most reliable methods, capable of killing most pathogens, but it requires time and resources that may not always be available. Disinfectant tablets, on the other hand, offer a portable and convenient solution, but it is important to follow the instructions carefully to ensure the treatment is effective.

Another emerging technology that is gaining traction is the use of ultraviolet rays. UV devices can neutralize microorganisms in water without altering its flavor or chemical composition, but require energy sources, such as batteries or solar power, which may not be available in all circumstances.

The key to effective purification is redundancy. Relying on just one method is not enough. You should have multiple solutions available, so you can address a wide range of scenarios. Security is never a given; it is the result of meticulous preparation and in-depth knowledge of the resources and tools at your disposal. In a world where certainties can vanish in an instant, knowing how to handle water safely is a skill you can't afford to neglect.

Your home should be a fortress, not only against visible threats, but also against invisible ones, such as water contamination. This is why knowing and implementing water filtration and purification methods is not just a precaution: it is a vital necessity. When safe water is in short supply, your level of preparedness will determine whether you can continue to provide a safe and secure environment for you and your loved ones.

CHAPTER 4:

Creating your own energy

Making your home a safe place is not just about physically protecting you from external threats, but also about creating a self-sustaining environment that can support you and your family in any circumstance. One of the most critical aspects of this self-sufficiency is the ability to generate and manage your own energy. In a world where energy access can be interrupted or compromised by natural events, social unrest or other emergencies, having control over your energy supply becomes essential. Creating your own energy is not just a matter of comfort, but of safety and survival. The first thing to understand is that energy is not a luxury, but a necessity. It powers every aspect of daily life, from meal preparation to lighting, heating and communication. Without a stable and reliable energy supply, even the most basic tasks become difficult, if not impossible. For this reason, it is essential to think ahead and develop an energy strategy that allows you to maintain self-sufficiency even in the event of prolonged interruptions to the electrical grid or other centralized forms of energy. One of the most effective solutions for creating your own energy is installing a solar panel system. Solar energy is an abundant, renewable resource that can be harnessed to power a wide range of home appliances. In addition to reducing your dependence on the electricity grid, solar energy allows you to have a clean and continuous source of energy, even during blackouts or emergencies.

However, to make the most of this technology, it is important to consider not only the installation of the panels, but also the storage capacity of the energy produced. Solar batteries, which store energy for use at night or on cloudy days, are a key component of any efficient solar system. Another crucial aspect to consider is the diversification of energy sources. Relying solely on a single source can leave you vulnerable in the event of a breakdown or adverse weather conditions. Integrating a fossil fuel-fired generator can provide an energy reserve in situations where there is not enough sun, such as during long storms or harsh winters. Even a small portable generator can make a difference, providing power for essential needs until conditions improve. However, it is important to use these generators with caution, ensuring that you always have sufficient fuel available and that you operate in a well-ventilated environment to avoid risks related to inhaling harmful gases. Energy efficiency is another key element. It's not enough to generate energy, you also need to know how to use it in the most efficient way possible. Thermal insulation of the home, the use of low-consumption appliances and the adoption of daily practices aimed at reducing waste can significantly extend the energy autonomy of your home. The more you can optimize your energy use, the less you will need, thus reducing your dependence on external sources and extending the life of your energy reserves.

Creating your own energy should not just be seen as an emergency measure, but as an investment in the future of your home and family. Energy security offers you not only the peace of mind of knowing that you are ready to face any situation, but also significant financial savings in the long term. Reducing your dependence on energy suppliers and fossil fuels not only protects you from price fluctuations and service interruptions, but also contributes to a more sustainable environment. That said, making your home a safe place requires careful planning and the ability to anticipate challenges you may face. Creating your energy is one of the most important aspects of this preparation. With the right technologies and a strategic approach, you can ensure that your home remains functional and safe in all circumstances, regardless of external conditions. Your energy is the foundation of your independence, and investing in it means investing in the safety and resilience of your home and family.

Energy requirements in case of emergency

Whether it's a natural disaster, a power grid outage, or another unexpected emergency, the ability to maintain a stable and reliable energy supply can mean the difference between overcoming the crisis and succumbing to hardship. Planning ahead for how to meet your home's energy needs is not just a matter of comfort, but of pure survival. In an emergency situation, access to energy quickly becomes a primary necessity. Without energy, you can't heat your home, store food, run medical equipment, or maintain communications. This is where preparation comes in: knowing what your essential energy requirements are and having a plan to meet them is key. Every home has different needs, so the first step is to evaluate what your priorities are. These could include maintaining a safe indoor temperature, powering communications devices, storing food and water, and maintaining basic lighting. The ability to identify these priorities and plan accordingly requires a detailed analysis of your home's energy consumption. This allows you to calculate how much energy is needed to keep essential functions running during an emergency.

Once you have a clear idea of your needs, you can begin to develop a plan to meet them. This could include installing solar panels to provide renewable energy, purchasing an emergency generator for situations where solar power is not sufficient, and storing fuel to ensure the generator can operate continuously . A crucial aspect of emergency energy management is efficiency. It's not enough to have an energy source; you need to be sure to use it as efficiently as possible to maximize its lifespan and usefulness. This means properly insulating your home to reduce heat loss in the winter and heat gain in the summer, choosing energy-efficient appliances and devices, and, if possible, centralizing energy-intensive activities at certain times of the day to reduce the overall load. Energy efficiency allows you to stretch available resources, thus increasing your ability to withstand a prolonged emergency. Another factor to consider is the resilience of your energy system. Emergencies often result in the disruption of traditional infrastructure, so it is essential that your energy system is able to operate independently.

This means having an energy source that is not dependent on the main electricity grid, such as a battery system powered by solar panels or a stand-alone generator. Additionally, regular maintenance of these systems is critical to ensuring they function optimally when you need them most. Don't wait until an emergency occurs to find that your generator isn't working or your solar panel batteries aren't charging properly. It is important to consider the duration of your ability to maintain energy. In some cases, emergencies can last days, weeks or even months. Having a sufficient supply of fuel, spare batteries, and a rotation plan for food and other resources will allow you to sustain you and your family for an extended period. Additionally, being able to adapt and modify your plan based on your circumstances is essential. Flexibility and the ability to respond quickly to changes in the situation are skills that are learned not just through planning, but through experience and mental preparation. Managing energy requirements in an emergency is an essential part of ensuring your home remains a safe place, regardless of what happens outside. Planning ahead, understanding your needs, optimizing efficiency and ensuring the resilience of your energy system are all crucial steps to ensuring that, whatever crisis you find yourself facing, you will be ready to support yourself and your loved ones. Your home should be your refuge, a place where you can face any challenge with confidence, knowing you have the resources needed to overcome any difficulty.

Off-grid power options: generators, wind and solar

When it comes to making your home a truly safe place, one of the most critical aspects is the ability to guarantee an energy supply independent of public networks. Living off the grid means you have total control of your energy, which is crucial in emergency situations, natural disasters or simply for those who want more autonomy. Off-grid power options, such as generators, wind, and solar, offer different solutions to meet your energy needs, each with its own unique benefits and considerations. Generators represent one of the most immediate and reliable solutions for guaranteeing energy independently. They run on fuels such as diesel, gasoline or propane, and are particularly useful when you need a fast, powerful energy source capable of powering essential appliances, medical equipment or heating systems. However, generators are not without their limitations. Their operation depends on the availability of fuel, which must be stored in sufficient quantities and carefully, as fuels can be dangerous if not handled correctly. Additionally, generators can be noisy and polluting, making them a temporary solution rather than a long-term strategy. Wind energy, on the other hand, offers a more sustainable, long-term approach to off-grid power. Installing wind turbines can generate enough electricity to cover a home's daily needs, especially in areas where the wind is constant and powerful.

Wind energy is clean and renewable, and once the turbines are installed, operating costs are relatively low. However, this type of energy depends on weather conditions; in calm periods, energy production can significantly reduce, requiring a backup energy source or storage system to ensure continuous supply. Solar energy is perhaps the most versatile and accessible of the off-grid power options. Solar panels can be installed on any building with sun exposure and, once in operation, offer a silent, emission-free source of energy. Solar energy is ideal for self-sufficiency, as panels can directly power home systems during the day, and with the use of storage batteries, electricity can be stored for use at night or on cloudy days. Like wind energy, solar energy also depends on environmental conditions, but with a good storage system and careful planning, it can cover a large part of a home's energy needs.

When choosing between these options, the primary consideration should be long-term reliability and the ability to ensure continuous power, regardless of external conditions. Often, a combination of these technologies offers the most robust solution: solar power for basic production during the day, a generator for emergencies or peak consumption, and wind power for continuous support, especially in particularly windy areas. This integrated strategy not only increases your home's energy independence, but also makes it more resistant to fluctuations and interruptions that can occur during emergencies. Regular maintenance is another crucial aspect. Any off-grid power system requires constant care to ensure it functions at its best when needed. Batteries need to be checked and replaced as needed, generators need to be started periodically to ensure they are operational, and wind turbines or solar panels need to be inspected to ensure there is no damage or obstructions that could reduce their efficiency. Making your home safe means preparing for any eventuality, and having a reliable power source is a key part of this preparation. No matter which option you choose, the important thing is that you are ready and that your home is equipped to face any challenge, safe in the knowledge that the lights will stay on, the water will be hot and the food will be preserved, no matter what happens outside. This is true security, the one that comes from preparation and the ability to be self-sufficient, in every situation.

Fuel safety and conservation issues

Fuel is an essential resource for powering generators, vehicles and other vital equipment during a crisis. However, its conservation presents a number of challenges and risks that must be addressed with extreme care. Managing fuel correctly is not just a matter of convenience, but of safety for you, your family and your home. The first point to consider concerns the nature of the fuel itself. Gasoline, diesel, propane and other fuels are highly flammable and can pose a significant hazard if not handled properly. Their conservation requires particular attention to environmental conditions. Fuel must be stored in containers approved and designed specifically for this purpose, in a well-ventilated area and away from sources of heat, sparks or flames.

Furthermore, it is essential to keep these reserves in a protected location, out of reach of children or animals, and preferably in an area separate from the main home to reduce the risk of house fires. Another critical aspect concerns the shelf life of the fuel. Different types of fuel have different shelf lives and tend to deteriorate over time. For example, petrol can begin to degrade after just three to six months if not treated with stabilizing additives, while diesel can last longer but is still subject to contamination by algae or sediment. Using expired or contaminated fuel can damage generator and vehicle engines, leaving you without power when you need it most. For this reason, it is essential to rotate fuel supplies regularly, using older ones before they deteriorate and replacing them with fresh new supplies. This way, you ensure that the fuel is always ready and in the best condition for use.

Safety is not limited to storage, but also includes transportation of fuel. When transporting fuel, it is essential to use approved, sealed containers, avoid overfilling them to allow the liquid to expand, and always keep them upright during transport. Any spillage should be treated immediately with specific absorbent materials, and the area should be well ventilated to avoid the accumulation of flammable vapours. Additionally, it is important to consider the possibility of an emergency situation where access to fuel may be limited or impossible. Your home security strategy should include not only fuel conservation, but also planning for energy conservation and efficient use of available resources. Minimizing fuel consumption through energy efficiency practices, such as using generators only when needed and adopting energy alternatives such as solar panels, can extend the life of your reserves and increase your self-sufficiency during a crisis. Finally, it is crucial to educate all members of your family about the risks associated with fuel and safety procedures.

Everyone needs to know how to handle and store fuel safely, how to respond to a spill, and what to do in the event of a fire. Having a well-defined emergency plan and conducting regular drills on how to respond to fuel-related incidents can make the difference between a minor inconvenience and a catastrophe. Fuel storage safety is an essential component of a safe and prepared home. Knowing the risks, taking the right prevention measures and maintaining a proactive attitude in managing your assets ensures that, whatever happens, you will be ready to protect your home and your family. Fuel is a precious resource, but like all precious things, it must be treated with respect and caution to ensure your safety and the safety of those around you.

CHAPTER 5:

Home defense techniques

Home security requires planning, attention to detail, and a prevention mindset. Every home has its weak points, and the key to making it safe is to identify them and strengthen them with determination. You must learn to observe your home with critical eyes, trying to anticipate any possible threat. This means not only thinking about how to protect main entrances, but also considering windows, garages, and any other access routes. Every corner of the house must be thought of as a potential vulnerability that requires an adequate solution.

Technology is a powerful tool, but it cannot replace intelligence and preparation. Installing sophisticated alarm systems, surveillance cameras and security lights is essential, but all of this must be integrated into an overall plan that also includes reacting to unexpected situations. It is not enough to rely on an alarm to feel safe: it is essential to know how to respond if the alarm sounds. Each family member must be aware of their role in an emergency, and these procedures must be practiced regularly, until they become second nature.

A safe house is a house where the element of surprise works in your favor. Intruders seek ease and anonymity. If you make your environment difficult to decipher and predictable, you exponentially increase your chances of avoiding an intrusion. This can be achieved through strategic lighting that eliminates shadowy areas around the home, the use of visible but not easily accessible security systems, and the creation of psychological barriers that discourage even the idea of attempting a break-in. Don't underestimate the impact of your presence: being visible, maintaining vigilant behavior and interacting with the local community all contribute to creating a safer environment.

The secret to effective home defense is maintaining a balance between visibility and unpredictability. Show that your home is protected, but without revealing all your cards. This can include the use of hidden cameras in addition to visible ones, alarm systems that do not follow fixed patterns, and adopting routines that never make it obvious when the house is empty. Security is never static: it must evolve over time, adapt to new threats and incorporate the latest technologies without becoming too dependent on them. Being prepared also means having a backup plan for when technology fails, because at the end of the day, the most valuable asset you have to protect your home is yourself.

Finally, remember that safety is not just about physical protection, but also about mental protection. Knowing that you and your family are ready to face any situation is the very heart of a safe home. No matter how sophisticated your security system, your peace of mind comes from knowing that when things get tough, you know exactly what to do. Preparation is your greatest weapon, and there are no shortcuts to getting it. It takes discipline, it takes practice, and it takes an unwavering dedication to protecting what matters most. With this mentality, your home will not just be a place to live, but a true bastion of security and serenity.

Defense of your home: windows, doors and exterior

The defense of your home does not begin and end with the front door, but involves every single detail: from the windows to the doors, up to the exterior of the property. Each element must be considered a line of defense, and every decision made must be considered as if the safety of your family depends on it, because it does exactly that. Windows often represent the most vulnerable point in a home, but they can become a bulwark of security if treated with the right attention. It is essential to install safety glass, such as tempered or laminated glass, which resists impacts better and makes forced entry more difficult. Also consider installing window locks that are easy for you to use, but a nightmare for anyone who tries to tamper with them from the outside. Grilles may seem like a drastic addition, but in high-risk areas they are a nearly impenetrable security measure. If you don't like the idea of turning your home into a fortress, there are more aesthetically pleasing solutions, such as security window films, which strengthen the glass without compromising views or natural light.

Doors are the first physical obstacle an intruder encounters, and they must be designed to resist. A sturdy door, preferably made of steel or solid wood, is just the beginning. The quality of the lock is equally crucial: opt for a European cylinder model, which is difficult to force and resistant to break-in attempts. But it's not just a question of materials; the often overlooked door hinge must be secure and reinforced, as a strong door is useless if it can be dismantled easily. Additionally, installing a magic eye or camera on your door can give you the ability to assess a potential threat before you even open the door. When it comes to secondary or garage doors, often considered less important, the same level of precaution must be taken, because a house is only as safe as its weakest point.

The exterior of your home is the first line of defense. A well-kept garden is not just an aesthetic issue, but a safety measure. Keeping trees and bushes pruned eliminates potential hiding places for anyone who wants to get close without being seen. Fences should be high enough and difficult to climb, but not too sturdy that someone can hide behind them. Lighting is a fundamental deterrent: a well-lit house is a less attractive house for those with bad intentions. Install motion sensor lights around all access points and also consider using timed lights to simulate someone's presence when you are away from home.

But the defense doesn't stop there. Having a video surveillance system, visible but well protected, adds an additional layer of security. The cameras not only record, but deter anyone considering intruding. And if you think that cameras alone are sufficient, you are wrong: they must be supported by an efficient alarm system, ready to go off at the slightest sign of intrusion. Everything visible on the outside must send a clear message: this house is protected, and entering will not be easy, nor without consequences. Home security requires a methodical and ongoing approach. It's not something you can set and then forget. It must be revised, updated, adapted to new threats and new technologies. Every family member must know what to do in an emergency, because the best defense is the one that has already been planned, rehearsed and internalized. Your home is not just the place where you live, it is the refuge where you should always feel safe. Make sure anyone who dares to test that confidence finds more than they bargained for.

Defense tools and techniques

It's a constant commitment, a mentality that requires you to always be one step ahead of those who might pose a threat. Home security is a complex set of tools and techniques that must work together to ensure that your home remains an inviolable place. You cannot afford to leave anything to chance, every detail must be taken care of with precision, every decision made must be aimed at creating an environment in which safety is a certainty, not an option.

The tools for the defense of your home are many and varied, but it is not enough to simply own them, you must know how to use them, you must understand how to integrate them into your overall strategy. An alarm system, for example, is not just a box that sounds when someone tries to enter. It must be configured to cover every possible access point, it must be connected to a monitoring system that can alert you immediately and, if possible, it must have the ability to function even in the event of a blackout or deactivation attempts. An effective alarm is one that not only warns you of danger, but does so in time to be able to react.

Security cameras are another crucial element. It's not just about recording what happens, but about letting anyone with bad intentions know that their every move is being watched. The simple presence of a camera can be a powerful deterrent, but it doesn't have to be the only one. The cameras must be positioned in strategic points, where they can cover blind spots, and must be connected to a system that allows you to monitor the situation in real time, even remotely. But remember, the cameras don't all have to be visible: a combination of visible and hidden cameras can seriously trouble anyone trying to breach your security.

In addition to technological tools, there are techniques you can adopt to make your home safer. One of the most effective is to create an unpredictable routine. Don't let a potential intruder study your movements, don't let your home ever seem empty for too long. Timed lights, timers to turn the TV or radio on and off, can simulate presence even when there is no one at home. These types of measures, while simple, can make the difference between an easy goal and one that is avoided.

In terms of physical defense, don't underestimate the importance of mechanical barriers. Doors and windows must be adequately reinforced, locks must be of high quality, resistant to burglary. If your home has a garden or outdoor space, make sure it is protected by adequate fencing and that access is controlled. Physical barriers don't have to turn your home into a fortress, but they should be strong enough to slow or deter an attempted intrusion, giving you time to react.

It's not enough to have the right tools, you have to know how to use them, you have to understand how to react if something goes wrong. Mental preparation is just as important as physical preparation. Each family member must know exactly what to do in an emergency, where to take refuge, how to communicate and how to act in a coordinated manner. Security is a team effort, and each member must be prepared to play their role. Drills aren't just for professionals, they're an effective way to make sure that if something ever happens, everyone knows exactly what to do without hesitation.

Defending your home is a commitment that requires discipline, focus and constant threat assessment. It's not something you can set and then forget, it needs to be an ongoing priority. Every time you introduce a new tool or technique, you need to consider how it fits into your overall strategy, how it can improve the security of your home. Only with a methodical approach, based on prevention and preparation, can you ensure that your home remains a safe place, a refuge where you can relax knowing that you have done everything possible to protect yourself and your family.

CHAPTER 6:

Interactions in times of crisis

In times of crisis, human interactions become one of the most critical elements in managing the security of your home. In an emergency situation, whether it is a natural disaster, a prolonged interruption of essential services or another form of crisis, the ability to communicate and interact with others can make the difference between maintaining control of situation or being overwhelmed by events. The interactions you have with your family members, neighbors and the wider community play a vital role in ensuring the safety of your home and protecting the people you care about. The first thing to understand is that a crisis tends to reveal the best and worst in people. In these moments, stress and fear can amplify emotional reactions and influence behaviors in unexpected ways. It is therefore essential to maintain a calm and rational attitude, especially when interacting with others. Your ability to remain composed and make thoughtful decisions will affect not only your personal safety, but also the safety of those around you. Words and actions in times of crisis must be guided by a mindset focused on solving problems and maintaining social cohesion. Communicating effectively is another crucial aspect. During a crisis, access to accurate and timely information becomes essential. It's important to establish clear communication channels with your family and neighbors in advance. These may include the use of radio devices, cell phones or other tools that function even in the event of a power outage or traditional communications infrastructure. Knowing who to contact and how to do so quickly can reduce uncertainty and facilitate coordination of actions needed to address the crisis.

Cooperation with neighbors is another key element. In times of crisis, the people who live near you can become a valuable resource. Sharing information, resources and mutual support increases the overall resilience of the community. However, it is crucial to establish clear boundaries early on and understand who among your neighbors is willing and able to collaborate constructively. Mutual trust must be built in times of peace, not when the pressure is at its highest. Forming local support networks, through neighborhood meetings or preparedness groups, can facilitate the creation of these relationships before a crisis hits. In addition to cooperation, it is important to be aware of potential threats from interactions with people outside your circle of trust. In situations of prolonged instability, the supply of limited resources can lead to tensions and conflicts. Protecting your home also means being prepared to handle these interactions confidently and decisively, avoiding unnecessary confrontations but being ready to defend what is yours if necessary. This requires not only mental preparation, but also awareness of the social dynamics and physical safety around your property.

The importance of psychological and moral support during a crisis should not be overlooked. Human interactions are not just about managing logistics, but also about maintaining hope and motivation. Being a point of reference for others, offering encouragement and emotional support, can strengthen collective morale and help everyone overcome the crisis with greater determination. This type of leadership, based on empathy and mutual respect, helps create a safer and more cohesive environment, where everyone feels valued and protected. In summary, interactions in times of crisis are complex and require preparation that goes beyond simple logistics. You need to develop a mindset that is ready to face challenges calmly, communicate effectively, build trusting relationships with those around you, and maintain a balance between cooperation and protection. The security of your home depends not only on the material resources you have accumulated, but also on the quality of relationships and interactions you are able to maintain during difficult times. Prepare to be a strength for yourself and others, and your home will become a safe haven, capable of withstanding any storm.

Reliable communication mechanisms put in place

In any crisis scenario, one of the key elements to ensuring the safety of your home is the ability to establish and maintain reliable communication mechanisms. Communication is the backbone of any operation, whether it is coordinating actions with your family members, receiving critical information from the outside, or interacting with neighbors and authorities. Without a robust communication system, you find yourself isolated, vulnerable, and unable to make informed decisions that can protect you and your loved ones. The first step to putting reliable communication mechanisms in place is to evaluate your specific needs and potential emergency situations you may face. Every crisis presents unique challenges, but the need to stay in touch with key people is a constant. Whether it's a natural disaster, a power outage, or a more localized threat situation, you need to be able to rely on a communications system that works even when traditional infrastructure fails. This requires a strategic approach and the use of technologies that ensure resilience and reliability. Cell phone communications are often the first medium people rely on, but during an emergency these networks can quickly become overloaded or completely down. For this reason, it is crucial to have alternatives available that do not depend on cellular networks. Two-way radios, such as Citizens Band (CB) or Amateur Radio (HAM), offer a more direct and independent level of communication, allowing you to maintain contact with those nearby without having to go through infrastructure centralized. These devices are relatively inexpensive and, when used correctly, can cover significant distances, especially in rural or suburban settings.

Another effective option is the use of satellite radios, which can ensure communications even in the absence of any terrestrial network. These devices are more expensive, but offer a near-universal connection, no matter where you are. Having a satellite radio as part of your communications plan gives you the security of being able to stay in touch with the outside world, even in the most extreme situations. However, as with any technological device, it is essential to familiarize yourself with how they work before an emergency situation arises, so that you can use them quickly and effectively when you need them. Maintaining effective communications during a crisis isn't just about technology, it's also about preparation and planning. It is essential that all members of your family are trained in the use of communications equipment and know the protocols to follow. Establishing in advance who to contact, in what order and through which channels, reduces chaos and increases efficiency in times of need.

This type of planning should include simple but clear communication codes, to quickly convey essential information without the risk of misunderstandings. It is important to have an emergency plan that also allows for the possibility of total loss of communication. Knowing what to do if you can't reach someone or if you suddenly find yourself isolated is just as important as having a functioning communication system. In these cases, preventive coordination and training are essential. Every person involved should know the emergency procedures, assembly points, and actions to take in the event of a communications breakdown. Finally, the importance of regularly testing your communications system should not be underestimated. Even the best plan is useless if the devices do not work properly or if the people involved are unable to use them under stress. Conducting periodic exercises and emergency simulations allows you to verify the functionality of your system and identify any gaps or problems that need to be resolved. Making your home a safe place inevitably involves creating reliable communication mechanisms. The ability to stay in touch, coordinate actions and receive accurate information is what will allow you to face any crisis with calm and control. With proper preparation, use of the right technologies and detailed planning, you can ensure that no matter what happens, you will always be able to communicate and protect those you love.

Digital security, radio and emergency broadcasts

In an increasingly connected and technology-dependent world, digital security and the ability to manage emergency broadcasts become essential elements to protect yourself and your family. The threat no longer comes only from physical forces or natural emergencies, but also from invisible attacks that can affect your network, your information and your ability to communicate at critical moments. Digital security starts with protecting your home networks. Every device connected to the Internet can become an access point for those who want to compromise your privacy or damage your systems. Securing your Wi-Fi network with strong, regularly updated passwords is just the first step. It is also essential to use high-quality firewalls and antivirus software to block intrusion attempts and detect suspicious activity. But digital security isn't just about devices. It's also about protecting your personal information.

Avoiding sharing sensitive details online, using two-factor authentication for your accounts, and being aware of digital scams can make a big difference in keeping your data and, in turn, your home safe. Alongside digital security, another pillar of protecting your home is the ability to effectively use emergency radios and broadcasts. In situations where traditional communications may fail, such as during a natural disaster or large-scale cyber attack, shortwave or two-way radios become vital tools for staying in touch with the outside world. These devices do not depend on traditional infrastructure and can operate in conditions where telephones and the Internet no longer work. Knowing how to use a HAM radio, for example, gives you access to a global communications network that can be critical for receiving real-time updates or calling for help. Furthermore, emergency broadcasts are not only a question of technology, but also of strategy. Knowing which frequencies to monitor, how to interpret emergency messages, and how to communicate effectively through these channels can make the difference between chaos and control.

Part of this preparation includes training. It's not enough to own a device; you need to be competent in its use. Knowing standard codes, call procedures, and how to convey clear and concise information during a crisis are skills that must be developed before the emergency arises. The combination of digital security and emergency broadcast management capabilities creates a resilient home environment, capable of withstanding not only physical threats but also the invisible ones that can affect your family in an increasingly interconnected world. These systems are not isolated; they must work together, as part of an overall security strategy that considers all possible risk scenarios.

Preparing your home also means ensuring that every component of your network, from your router to emergency communications devices, is secure, up-to-date and ready to use. In this context, it is essential to never underestimate the importance of preparation. Emergencies don't announce their arrival, and when they strike, there's no time to learn on the job. You need to be ready, your equipment needs to be ready, and everyone in your family needs to know what to do, how to communicate, and how to maintain information security. Remember that in a crisis, your actions and preparation are what will keep you safe. In summary, the security of your home today cannot ignore the protection of your digital communications and the ability to manage emergency broadcasts. Every device, every network, and every communication method must be part of a coherent strategy, designed to protect you and your loved ones from a broad spectrum of threats. With the right preparation, you can turn your home into a bastion of security, ready to resist not only physical attacks, but also the digital ones that are becoming increasingly common in the modern world.

Keep up to date without Internet access

The sudden absence of information can create confusion, uncertainty and, above all, jeopardize your ability to make informed decisions. Preparing for these situations requires a strategic mindset and adopting alternative methods to stay informed, even when the web is down. The first thing to understand is that information is power. In an emergency situation, knowing what is happening around you allows you to anticipate problems, react quickly and protect yourself and your loved ones. Without the Internet, it is essential to resort to reliable and alternative sources of information. Radio, especially AM and FM broadcast stations, becomes an indispensable tool. Many local and national stations broadcast news, weather updates and emergency alerts that can be essential to understanding the evolving situation.

Owning a battery-operated or hand-cranked radio that isn't dependent on electricity or the grid is a critical step in ensuring you have access to this information. Shortwave and two-way radios, such as CB radios or amateur radios (HAM), offer additional advantages. These devices not only allow you to listen to local broadcasts, but also allow you to communicate directly with other people who are in the same situation as you. Amateur radio networks, in particular, are often well organized and can provide vital information, especially in scenarios where traditional communications are disrupted. Training and practice in using these tools is essential: you need to be sure you know how to use them correctly and that you know the right frequencies and channels to monitor in an emergency. In addition to radio, the traditional press can also play an important role. Having paper maps, emergency manuals and how-to guides available allows you to navigate information without having to rely on electronic devices.

These materials can be pre-planned and researched in advance, ensuring that you are prepared to handle an information blackout situation. Information on local resources, emergency contact numbers and standard operating procedures must be easily accessible in physical format. The importance of human relationships in gathering information should not be underestimated. In the absence of the Internet, direct communication with neighbors, local authorities and community networks becomes a primary source of news and updates. Creating and maintaining good relationships with those around you not only strengthens your sense of community, but also ensures a continuous flow of information, exchanging news and resources that could be crucial to your safety. Being an active part of a local network allows you to obtain real-time information and coordinate with others to deal with emergencies more effectively. You cannot afford to remain isolated or ignorant when it comes to protecting your home and family. With the right preparation and a strategic approach, you can ensure you always have access to the information you need, regardless of external circumstances. This will not only give you the confidence to deal with any crisis, but will also put you in a position of strength and control, ready to make quick and effective decisions to keep your home a safe place.

CHAPTER 7:

Preparation for healthcare and basic first aid

Living in a world where critical situations can arise unexpectedly requires a well-defined strategy, which includes not only the physical protection of your home, but also preparation to deal with urgent medical events. Home defense begins with knowing the terrain. I'm not just talking about the interior layout, but everything that surrounds your property. Every access point, every blind spot, every possible vulnerability must be assessed and fortified. A sturdy door, windows with secure locks and an effective alarm system are essential, but what really matters is mental alertness and the ability to react quickly.

Security doesn't stop at protection against intrusions. It is essential to be prepared to handle emergency health situations. This means having a well-stocked first aid kit available, but above all knowing how to use it. Knowledge of first aid is not an option; it is a duty. Knowing how to stop a bleeding, treat a fracture or manage a choking situation can mean the difference between life and death. Time is a critical factor, and often the ambulance may not arrive in time. Health preparation is not limited to basic first aid. In an unpredictable world, knowing how to manage a more complex health crisis is equally important. Knowing the signs of a heart attack, knowing how to stabilize a patient until help arrives, or having basic skills in managing more serious injuries are skills that everyone should acquire. A simple training course can equip you with the skills necessary to deal with situations that require decisive and targeted interventions.

A safe home is the reflection of a prepared mentality. Security must be planned, tested and continuously improved. Each family member must know emergency procedures, know where first aid tools are located and how to use them. The drills aren't just for professionals; they must become part of the domestic routine, so that in case of need, intervention is automatic and without hesitation.

Keep essential medicines on hand

When it comes to protecting what really matters, being prepared to handle a health emergency is as critical as having strong locks on your doors. Time is a crucial factor in many critical situations, and having essential medications on hand can mean the difference between life and death. It is important that your home is stocked with all those basic medications that can help immediately manage a wide range of emergencies. I'm talking about painkillers, antihistamines, heart and blood pressure medications, insulin for those who need it and remedies for the most common conditions such as fever, headaches and gastrointestinal problems. However, it is not enough just to have these drugs at home; It's essential that they are easily accessible and well organized, so you can find and use them quickly when needed. Every second counts when someone has a heart attack, a severe allergic reaction, or a diabetic crisis, and you can't afford to waste time looking for what you need.

Preparation goes beyond simply storing medicines. You need to make sure everyone in the house knows where they are stored and how to use them properly. There is no room for improvisation in a crisis. The instructions should be clear, and if necessary, you should consider keeping a list of medications with directions for use clearly visible. Even a simple mistake, such as administering the wrong dose or confusing one drug with another, can have serious consequences. Training and awareness are an integral part of home security.

In addition, you need to be aware of the storage conditions of the drugs. Many medicines can lose their effectiveness if they are not stored correctly. Keep them in a cool, dry place, out of direct sunlight and out of reach of children. Check expiration dates regularly and replace expired medications immediately. An ineffective drug in an emergency situation is worse than having nothing at all. Preparing also means planning for the most serious situations. If anyone in your home has a medical condition that may require urgent treatment, make sure you have sufficient supplies of the necessary medications and know the procedures for administering them. You can't afford to depend solely on getting to a pharmacy or waiting for help to arrive. Each family member must be informed and educated on how to handle the situation until a medical professional arrives. Make preparation your most powerful weapon, and you'll always be one step ahead of danger.

Create a complete first aid kit

A first aid kit is not just a box with some bandages and disinfectant. It should be a resource you can count on in any situation, whether it's a superficial cut or a more serious emergency. The key is thoroughness. The kit should contain everything you might need to manage wounds, burns, trauma, breathing problems and any other situation that requires immediate intervention. Each element of the kit must be chosen carefully, thinking not only of the most common emergencies, but also of the less probable but potentially lethal ones. But having a complete kit is not enough. You have to know where it is, how to use it, and you have to make sure everyone in the house is equally knowledgeable. The kit should be placed in an easily accessible place, not hidden in a corner or in a cupboard behind a thousand other things. The location must be strategic, chosen so that it can be reached quickly by anyone who needs it, without having to think about it too much. Every family member must know exactly where it is and what it contains, because in a critical situation there is no time to look for or read labels.

Don't underestimate the importance of training. Knowing the contents of the kit is helpful, but knowing how to use each tool and medicine correctly is essential. It's one thing to know that you have an elastic bandage, it's another to know how to apply it correctly to stop a hemorrhage. It's one thing to have a defibrillator at hand, it's another to know how to use it in those crucial seconds that can save a life. For this reason, it is essential that you invest time in learning first aid techniques, and that this learning is extended to everyone who lives with you.

Also remember that your kit is not something you can assemble once and forget about. It must be maintained and updated regularly. Medicines expire, bandages deteriorate, and your needs may change over time. Periodically check the contents of the kit, replace what is expired or worn, and adapt the contents to the specific needs of your family. If someone develops a medical condition that requires specific medications or equipment, make sure these are included in the kit and that everyone knows how to use them. Ultimately, the safety of your home depends on how prepared you are to handle the unexpected. You can't afford to be caught off guard. Your first aid kit should be an extension of your determination to protect what you love, a symbol that you are ready to face any challenge with determination and competence. Don't leave anything to chance, because safety is not just a question of prevention, but also of preparation. Always be ready, and you will always be safe.

Medical knowledge that everyone should have

Medical knowledge is not something reserved only for health professionals. Everyone should be able to recognize the signs of a medical crisis and know how to intervene. For example, understanding when an injury requires immediate treatment or how to manage a person who is having a heart attack is crucial. In these moments, there is no time for panic or uncertainty. You have to know exactly what to do, and you have to do it quickly and with confidence. This type of preparation allows you to take control of the situation, rather than being overwhelmed by it. A crucial aspect of home safety is knowing how to handle common medical emergencies such as burns, deep wounds, severe allergic reactions, or choking incidents. These are situations that can happen to anyone at any time, and you must be prepared to intervene with confidence. But medical knowledge is not limited to wound management. You should also know how to recognize the signs of more serious conditions such as a stroke or internal bleeding, where every second counts.

First aid training should be an essential part of your preparation. It's not enough to have a well-stocked first aid kit; you need to know how to use it effectively. A first aid kit without the right knowledge is like a weapon without training: useless, and in some cases, even dangerous. Invest the time necessary to learn the basic techniques and make sure everyone in the house has the same skills. In an emergency situation, anyone could be called upon to intervene. In addition to practical skills, it is important that you develop a problem-solving mindset. Medical emergencies never unfold as expected. You must be able to adapt, make quick decisions and remain calm under pressure. Your ability to manage stress and act with determination can be the difference between a successfully handled incident and a tragedy. This mentality is not innate; it is built through preparation, experience and constant updating of your skills. Finally, don't forget the importance of communication. In a crisis, the ability to coordinate efforts and communicate clearly with those around you is crucial. Whether it's instructing a family member on how to perform first aid or quickly explaining a situation to medical personnel over the phone, your ability to convey information clearly and effectively can save lives.

CHAPTER 8

Maintain hygiene and sanitation

When resources are limited and access to essential supplies may be compromised, sanitation and hygiene become key priorities. Maintaining high hygiene standards is not just a matter of comfort, but a vital necessity to prevent illness, protect your health and that of your family, and ensure your home remains a safe haven. Maintaining hygiene in challenging conditions requires planning and preparation. In crisis situations, clean water may become scarce, making it essential not only to accumulate water reserves, but also to adopt purification and conservation techniques. Water is not only necessary for drinking, but also for cooking, washing and keeping living spaces clean. Without proper water management, the risk of infection and disease increases exponentially. It is therefore crucial to have access to water purification methods, such as filters, disinfectant tablets or boiling, to ensure that the water available is safe for all uses.

Regular home cleaning becomes even more important during a crisis, as the buildup of dirt and waste can quickly turn into a health hazard. Keeping surfaces, utensils and common areas disinfected reduces the risk of spreading bacteria and viruses, which can thrive in neglected environments. Cleaning products, such as disinfectants, bleach and soap, should be part of your essential supplies, as they are essential tools for maintaining a safe environment. Even in the absence of commercial products, it is possible to create effective cleaning solutions with basic materials, such as vinegar and baking soda, which can be used to sanitize everyday surfaces and objects. Personal hygiene is equally crucial. In an emergency situation, where access to personal hygiene facilities may be limited, it is necessary to develop routines that allow cleanliness to be maintained even in less than ideal conditions.

Washing your hands regularly with soap and water, especially before eating or after touching dirty objects, is one of the most effective practices for preventing the spread of diseases. If water is scarce, the use of alcohol-based hand sanitizers may offer a viable alternative. The management of domestic waste, including organic and human waste, must also be approached carefully to avoid contamination and the proliferation of parasites. It is also important to be prepared to handle any illnesses or infections that may arise. Having a well-stocked first aid kit, with basic medications, bandages and disinfectants, is essential. The ability to recognize the symptoms of common infections and illnesses, and treat them promptly, can prevent complications and reduce the need to resort to outside medical facilities, which may be overwhelmed or unavailable during a crisis. In conclusion, maintaining hygiene and sanitation in an emergency situation requires a combination of preparation, resources and awareness.

Your home must be not only a safe place from external threats, but also a healthy environment where you and your family can live safely. Every small gesture of cleaning and hygiene, if done with consistency and attention, helps to create a barrier against diseases and infections, protecting the health of everyone who lives under your roof. Preparing for these eventualities is not just a matter of having the right materials, but of developing a careful and disciplined mindset, ready to maintain high standards of hygiene even in the most difficult circumstances. This type of preparation allows you to face any situation with the certainty that you are ready to protect yourself and your loved ones, not only from visible threats, but also from invisible ones that can endanger your health and safety.

Waste management solutions for extended settlements

In crisis or emergency situations, the uncontrolled accumulation of waste can quickly turn into a serious health problem, leading to diseases, infestations and a general compromise of the liveability of the home environment. Managing waste safely and efficiently is not just a matter of order, but an essential component to maintaining the health and safety of those who live in your home. The first thing to understand is that, in a context of prolonged settlement, waste production does not stop. Indeed, it may increase, especially if normal waste collection and disposal facilities are interrupted or unavailable. For this reason, it is essential to have a clear and practical plan for waste management. This plan must be flexible, adaptable to available resources and capable of preventing problems before they become insurmountable. Effective waste management starts with minimizing the amount of waste produced. This can be achieved through practices such as reusing materials, composting organic waste and reducing unnecessary packaging.

Every object that does not become waste is a step forward in overall management. In a context where the public waste collection service is compromised, it is essential to establish safe methods for the disposal of solid and liquid waste. Organic waste can be managed through composting, a practice that not only reduces the volume of waste but also produces natural fertilizer that can be used to grow plants, thus increasing the sustainability of the settlement. However, composting must be handled with care to avoid odor or infestation problems. Choosing a suitable location, aeration and moisture management are key factors for safe and effective composting. Human waste management requires special attention. In the absence of functioning sewage systems, alternative solutions must be implemented to ensure that human waste does not contaminate drinking water or the surrounding environment. Compostable latrines, which isolate waste and allow it to decompose safely, can be an effective solution.

These systems reduce the risk of waterborne diseases and prevent soil and groundwater pollution. In more extreme situations, it may be necessary to resort to improvised latrines, but even in these cases it is essential to follow rigorous hygiene practices to minimize health risks. Non-organic waste, such as plastic, metal and glass, must be managed to avoid accumulations that could become dangerous or polluting. If it is not possible to recycle or reuse these materials, they should be stored safely, away from inhabited areas, until appropriate disposal is possible. It is important to keep these storage areas well organized and separated from organic waste to prevent contamination and to facilitate any future disposal operations.

Waste management for extended settlements requires a combination of practicality, ingenuity and discipline. Every member of your family needs to be involved and educated on correct practices to ensure everyone is aware of the risks and solutions available. Cooperation and awareness are essential to maintaining a safe and healthy home environment. It's not just about keeping your home clean, it's about ensuring your home remains a safe and sustainable haven, even in the most challenging circumstances. Ultimately, effective waste management in a prolonged settlement is not a task to be taken lightly. It requires planning, adequate resources, and an ongoing commitment to ensuring that your home remains a place where you can live safely and healthily. With the right measures, you can prevent waste from becoming a threat and turn waste management challenges into an opportunity to improve the sustainability and resilience of your home. This is a critical aspect of any home security plan and one that cannot be overlooked if you want to be truly prepared to deal with any situation.

Maintain personal hygiene: Keep yourself clean and healthy

Maintaining personal hygiene is one of the most crucial aspects of ensuring safety and health within your home, especially in crisis situations. When resources are scarce and normal routines are disrupted, maintaining high standards of personal cleanliness becomes a vital necessity, not only to prevent illness, but also to sustain morale and maintain a sense of normality. Personal hygiene, in fact, is much more than a question of comfort: it is a fundamental line of defense against infections and diseases, which can spread rapidly when hygienic conditions deteriorate. In a context where access to clean water may be limited, it is essential to adopt practices that maximize the effectiveness of available resources. Washing your hands regularly is the first line of defense against the transmission of germs and bacteria. Even in the absence of abundant running water, good hand hygiene can be maintained by using alternative methods, such as alcohol-based sanitizers or sanitary wipes. These tools can be extremely effective in reducing bacterial load, especially when used correctly and consistently. Skin care is another essential element of personal hygiene.

The skin is the body's first barrier against infections, and keeping it clean and healthy is essential. Even in the absence of daily showers, you can use damp cloths or towels moistened with warm water to clean the areas of the body most prone to sweating, such as armpits, groin and feet. This not only helps prevent skin irritation and infections, but also helps maintain a sense of freshness and general well-being, which is vital for morale in difficult situations. Maintaining good oral hygiene is equally crucial. Even in situations where it may not be possible to follow your normal twice-daily toothbrushing routine, it is essential to do everything you can to keep your mouth clean. If water is in short supply, use a small amount to rinse your mouth or, if necessary, chew sugarless gum which stimulates saliva production, helping to maintain the bacterial balance in your mouth. We must not overlook the importance of hair and nail care. While they may seem like minor issues, keeping your hair clean and nails short reduces the risk of infections. Dirty hair and long nails can hold dirt, bacteria, and other pathogens that can easily transfer to the rest of the body. Furthermore, taking care of these details helps maintain a sense of normality and dignity, important psychological elements in any emergency situation. In prolonged situations, where access to sanitation may be limited, management of body waste becomes a crucial component in maintaining personal hygiene and preventing the spread of disease. It is crucial to establish safe waste disposal practices and ensure that every person in your home is trained to strictly follow them. These practices not only protect individual health, but also help keep the home environment safe for everyone.

Maintaining personal hygiene is closely linked to psychological well-being. Feeling clean and fresh helps keep your spirits high, reducing the stress and anxiety that can arise from a crisis situation. It is important to recognize that personal hygiene is a form of self-care, a routine that, if maintained, can provide a sense of control and normality even in the most difficult circumstances. Ultimately, maintaining personal hygiene in an emergency situation requires discipline, ingenuity and a strong sense of responsibility. Every little action counts, and your commitment to personal hygiene not only protects your health, but helps keep your home a safe haven for all its inhabitants. Even when resources are limited, maintaining high standards of cleanliness and hygiene is possible, and will make the difference between getting through a crisis healthy or facing avoidable complications. With proper preparation and a solution-oriented mindset, you can ensure that no matter the situation, you will be able to stay clean, healthy, and ready to take on any challenge.

Avoid illnesses in confined spaces

When living in confined spaces, such as during an emergency situation where the entire family is forced to remain confined indoors, maintaining health and preventing the spread of disease becomes a top priority. In these conditions, the population density increases, and with it the risk of contagion and the onset of diseases. To ensure that your home remains a safe place, it is essential to take specific measures to prevent viruses and bacteria from finding breeding ground in cramped environments. The first line of defense against illness in confined spaces is personal and environmental hygiene. When people live in close proximity, the chance of transmitting germs through direct contact or contaminated surfaces increases exponentially. Maintaining a rigorous hand cleaning routine and making sure everyone follows this rule is crucial. Washing your hands regularly with soap and water, especially before eating or after touching common surfaces, can make a big difference in preventing the spread of pathogens. In the absence of running water, the use of alcohol-based disinfectants is an effective alternative that must always be at hand. Furthermore, the hygiene of the surrounding environment is equally crucial. Surfaces that are touched frequently, such as door handles, light switches and faucets, should be cleaned and disinfected regularly. This reduces the chance of germs accumulating and spreading among family members. Using appropriate disinfectant products and following the instructions for their use ensures that sanitization is effective and continuous.

Air management within the confined space is another key factor in preventing disease. Stagnant and recycled air can become a vector for the spread of respiratory diseases, especially in closed and crowded environments. Ensuring good ventilation, by opening windows when possible or using air purifiers with HEPA filters, helps keep the air clean and reduce the concentration of infectious particles in the environment. Even simple measures such as avoiding smoking indoors or using chemicals that release toxic fumes can significantly improve air quality. In an emergency setting, where medical resources may be limited, it is essential to carefully monitor the health of all family members. Recognizing the symptoms of infectious diseases early, such as fever, cough or general malaise, allows you to isolate the patient and prevent the spread of the infection. Timely isolation in a separate space, even if small, can be crucial to containing the disease and protecting the rest of the family.

Waste management in confined spaces is another element that should not be underestimated. Waste, especially organic and medical waste, can quickly become a breeding ground for bacteria and parasites if it is not disposed of correctly. Having an efficient system for the collection and removal of waste, which ensures that it does not remain within the inhabited space for long periods, is essential. This may include the use of airtight waste bags and their regular disposal in appropriate places. Finally, maintaining a positive mood and managing stress is equally important for overall health. Prolonged stress can weaken the immune system, making people more susceptible to disease. Creating an environment that promotes mental well-being, even in confined spaces, is possible through recreational activities, exercise and open communication between family members. The awareness that everyone has an active role in maintaining everyone's health helps strengthen cohesion and reduce the risk of disease. In summary, preventing disease in confined spaces requires a combination of discipline, awareness and preparation.

Every gesture, from cleaning hands to ventilating rooms, contributes to creating a safe and protected environment for everyone. In emergency situations, your home must remain not only a physical refuge, but also a bastion of health, where prevention is the key to overcoming any challenge. With constant attention and a proactive approach, you can ensure that your family remains healthy and safe, regardless of external difficulties.

CHAPTER 9:

Crucial skills for survival

Security is a mentality, a way of life that goes beyond simple physical protection. It is the ability to anticipate, prevent and react effectively to any threat, turning your home into a fortress that not only resists attacks, but prevents them. Imagine your home as an operational base, a place where every detail is designed to guarantee the maximum possible protection. To get started, you need to know your environment better than anyone else. The first rule is situational awareness. You need to know who and what is moving around your home, and this requires an in-depth study of your habits and those of your neighborhood. Knowing the weak points of your property is essential: windows without adequate protection, doors that do not close properly or blind spots that can be exploited by intruders. Every entrance and exit must be guarded, and every vulnerability must be fortified.

Investing in a good security system is essential. Well-placed cameras, motion sensor lights and a reliable alarm can make the difference between a failed attempt and a successful intrusion. But equipment alone is not enough. You need to train yourself and your family on how to react in an emergency. Everyone needs to know what to do if the alarm goes off, if someone approaches the house or if there is a power outage. Readiness is everything. You also need to consider the internal layout of your home. Arranging furniture and organizing spaces so that you always have a clear escape route is crucial. In an emergency situation, every second counts, and having to navigate unnecessary obstacles can be costly. At the same time, make sure there are safe shelters inside the house, where you can barricade yourself if necessary. These spaces must be equipped with everything necessary to resist for a prolonged period: food, water, communication tools and, if possible, means of defense.

Don't underestimate the importance of a good communication network. In a crisis situation, being able to count on friends, neighbors or family members who can intervene quickly can be crucial. Establish clear signals to communicate emergency situations, and make sure everyone who lives with you knows how to use them. Finally, the most important thing is mentality. Being ready does not mean living in fear, but living with the awareness that safety is an ongoing process, which requires vigilance, preparation and a proactive attitude. You must always be ready to defend what is yours, and this requires discipline and determination. Remember, a safe home is the result of a series of conscious choices, made every day.

Light a fire and preserve the heat. Basic methods of self-defense

Knowing how to start a fire and maintain heat becomes essential to survival, not only for comfort, but also for protection. The ability to light a fire, using natural or makeshift materials, is one of the most basic but also most crucial skills. Whether you're using a lighter, a flint, or more primitive methods like friction, the goal is the same: generate heat and maintain it. It's not enough to know how to make a fire; you also need to know how to do it safely, how to protect it from wind and moisture, and how to constantly power it to ensure it doesn't shut down when needed.

Preserving heat isn't just about fire. In difficult situations, every available resource must be exploited. Insulation is one of the keys to maintaining body heat and can be achieved with simple materials you may have on hand. Blankets, warm clothing, leaves or even paper can make the difference between a freezing night and a manageable night. Knowing how to create a barrier between you and the cold is a lesson that can also be applied to protecting your home: every crack, every weak point in your home represents an opportunity for the cold, or worse, for an intruder.

Moving on to self-defense, this is another critical component of safety, both personal and home. You don't have to be a martial arts expert to defend yourself effectively, but you do need to be clear on some basic principles. Environmental awareness is key – you need to know what's around you and have a plan in case of an emergency. In a physical confrontation situation, the goal is not to win, but to survive and get away from the threat as quickly as possible. Basic self-defense techniques include striking vulnerable points on your opponent's body, such as the eyes, throat, and knees, to create an opening for you to escape. Strength is not always necessary, but precision and determination are.

By combining these skills, from heat management to self-defense, you can create a home environment that not only protects you from outside threats but allows you to be self-sufficient in an emergency. A safe home is a prepared home, and preparation begins with knowledge and the ability to apply it in any situation. It's not just about building walls or installing alarms, it's about having the skills to face the unknown with confidence and determination. This is the kind of mindset that can make all the difference when it comes to protecting yourself and your home.

Find your way and send signals for help

Knowing how to find your way and sending signals for help are crucial skills that can make the difference between a quick resolution to an emergency and a situation that quickly escalates. In an urban setting, knowing your location and knowing how to navigate may seem trivial, but in emergency situations, such as a blackout, a natural disaster or an intrusion into your home, your ability to orient yourself quickly and accurately becomes critical. You must have a clear understanding of the layout of your neighborhood, escape routes, and safe spots where you can find shelter or assistance. If you find yourself in an unfamiliar area, having a physical map on hand or being able to use a GPS device efficiently is vital. However, you also have to be prepared to face the possibility that the technology fails, and in that case, knowing how to orient yourself using natural or urban landmarks becomes essential. Learning to read the sun, the stars, or simply recognize visual cues such as traffic flow or wind direction can help you stay in the right direction and avoid dangerous situations.

Sending signals for help is another crucial aspect of safety that should not be underestimated. If you find yourself in a situation where you are stranded or unable to leave your home, knowing how to communicate your need for assistance can save your life. Traditional methods, such as using a whistle or lighting a fire with lots of smoke, are effective in remote settings, but in an urban or suburban setting, you may have to rely on other tools. Flashlights with SOS functions, mirrors to reflect sunlight, or even using colored materials to attract attention can be effective ways to signal your location. Using simple visual cues, such as flags or banners, can also be helpful in indicating your presence to those who may be able to help you. You must also be able to remain calm and send clear and repeated signals, to avoid them being ignored or misunderstood.

These skills are not only useful in extreme situations, but can also make your daily life safer. For example, if you notice someone following your routine or loitering suspiciously near your home, knowing how to find a safe escape route or knowing how to send a discreet signal to a neighbor or law enforcement can prevent a direct confrontation. Security is not just a matter of physical strength or barriers; it is a question of intelligence, preparation and mental alertness. Ultimately, a safe home is one where everyone knows what to do, how to move and how to ask for help, regardless of the circumstances.

CHAPTER 10:

Psychological resilience

There is an equally important, often overlooked element that is essential to the true safety of your home: psychological resilience. In a crisis or emergency situation, the ability to stay calm, adapt to changes and deal with stress with clarity is what can make the difference between weathering the storm or being overwhelmed by it. Psychological resilience is the pillar on which the ability to respond effectively to difficulties, to remain focused and to make rational decisions even under pressure rests.

Psychological resilience is not an innate quality that only some possess. It is a skill that can be developed and strengthened with time and experience. It starts with self-awareness, the ability to recognize your limits and accept that, in certain circumstances, it is normal to feel overwhelmed. However, resilience doesn't stop at acceptance; requires a proactive attitude towards difficulties. It means turning uncertainty and fear into opportunities to grow, to learn something new and to become stronger. This process begins long before the emergency arises. Mentally training yourself to deal with stress, through breathing techniques, meditation, or even simply imagining difficult scenarios and how you would deal with them, can prepare you to respond calmly and in control when things get really messy.

Another key aspect of psychological resilience is social support. Even the strongest among us need a support network to rely on. In times of crisis, having trusted people to talk to, to share concerns and responsibilities, is essential. This not only relieves the mental load, but also allows you to see things from a different, often clearer and more rational perspective. Cultivating these relationships, building a community based on trust and mutual support, strengthens resilience not only on an individual level, but collectively. In a family context, psychological resilience is especially important because your ability to remain calm and centered will have a direct effect on those around you, particularly children and older adults, who tend to mirror the emotions of the adults around them. Part of psychological resilience is also the ability to be flexible. Emergencies rarely follow a script, and the ability to adapt to new situations, revise plans, and improvise is crucial.

This doesn't mean lacking plans or preparation, but rather having the mental strength to not rigidly cling to an idea when circumstances change. Mental flexibility allows you to see the options before you, find creative solutions, and make decisions that, in other situations, may not be the most obvious. Finally, psychological resilience is also built on hope and realistic optimism. Even in the darkest situations, it is important to maintain a positive outlook and believe that things can get better. This doesn't mean ignoring problems or pretending they don't exist, but rather approaching them with the belief that you can overcome them. Hope is a powerful motivating force; it pushes you to keep fighting even when all seems lost. Cultivating a positive attitude, maintaining faith in yourself and others, gives you the strength to persevere and find the light at the end of the tunnel. In conclusion, making your home a safe place goes beyond physical and practical measures. It is a commitment that also requires building psychological resilience that allows you to face adversity with strength and determination. Preparing yourself mentally, building strong relationships, and cultivating a flexible, positive attitude are critical steps to ensuring that no matter what challenges you face, you'll be able to keep your home a safe haven, not just physically, but emotionally as well. Psychological resilience is the heart of security, because it is what allows you to face any situation with the certainty that, in the end, you will overcome this too.

How to manage stress and isolation

In a crisis situation, when your home becomes the refuge where you can protect yourself and your family, one of the most difficult aspects to deal with is managing stress and isolation. These two factors, if not managed correctly, can erode your ability to make effective decisions, undermine your mental health, and turn a supposedly safe environment into a source of further tension. Understanding how to cope with and mitigate stress, along with how to manage feelings of isolation, is critical to keeping your home a safe and stable place during challenging times. Stress, in emergency situations, is a natural response. The body and mind enter a state of alert to deal with perceived threats, which can be helpful in short periods, but if stress persists, it can have deleterious effects on mental and physical health. The first thing to do is recognize stress for what it is: a response of the body to external pressures. It's not about eliminating stress completely, but managing it so that it doesn't take control of your actions and thoughts. One of the most effective techniques is maintaining a daily routine. Even in isolation, having set times for sleep, meals, exercise and housework helps create a sense of normality and predictability that can reduce anxiety and feelings of overwhelm. Physical movement, even as simple as doing bodyweight exercises or a short walk if space allows, is a powerful tool for relieving stress.

Physical activity releases endorphins, the so-called "happy hormones", which improve mood and reduce accumulated tension. Isolation, however, presents a different but equally significant challenge. Being isolated, whether physically from others or simply feeling cut off from the world, can lead to feelings of loneliness, depression and a general sense of alienation. In these moments, it is vital to maintain contact with the outside world, even if only virtually. Modern technologies offer many tools for staying in touch with friends and family, and these connections are crucial for maintaining high spirits and feeling part of a community, even from a distance. Regular communication not only combats isolation, but allows you to share your concerns, easing any emotional burden that may be accumulating. At the same time, it is also important to learn how to be alone without feeling lonely. This requires a certain psychological resilience, an ability to find comfort and strength within oneself. Meditation, reading, listening to music or even engaging in creative projects are activities that can help keep the mind occupied in a positive way, promoting a feeling of calm and inner well-being. Learning to manage isolation also means knowing how to cultivate your own mental space, transforming moments of solitude into opportunities to reflect, grow and regenerate. Finally, in prolonged situations of stress and isolation, it is essential to monitor signs of emotional overload, both in yourself and in other members of your family.

Symptoms such as irritability, excessive tiredness, difficulty sleeping or concentrating may be signs that stress is exceeding your tolerance level. In these cases, it's crucial to intervene promptly, finding ways to take the pressure off, whether that's taking a break, delegating tasks, or simply talking openly about your concerns. The ability to recognize these signs and proactively address them is a key component to maintaining a safe and healthy home environment. Ultimately, managing stress and isolation during a crisis is a complex challenge, but with the right strategies, you can stay calm, protect your mental health, and continue to function effectively. Your home must be a refuge not only physically, but also mentally, where stress is managed and isolation does not become an insurmountable obstacle. With a resilient mindset, a well-structured routine and a strong support network, you can turn even the most difficult moments into an opportunity to strengthen your ability to resist and overcome adversity. The key is to remain proactive, aware and ready to adapt, knowing that your inner strength is the true pillar of your home's security.

Develop mental strength

There is a less visible, but equally crucial element that plays a critical role in your ability to protect yourself and your family: mental strength. Mental strength is what allows you to face challenges with resilience, stay calm under pressure, and make clear, decisive decisions when it matters most. Developing it is not a luxury, but a necessity for anyone who wants to be truly prepared to face difficult situations. Mental strength is not something that is acquired overnight. It is the result of constant commitment, of training that involves both the mind and the body. It starts with the ability to deal with uncertainty. In crisis situations, things rarely go as planned. Unexpected events are part of the game, and your ability to quickly adapt to changing circumstances is what will determine how effectively you handle difficulties. This requires accepting uncertainty not as an enemy, but as an inevitable part of life. Training your mind to remain flexible, not to tense up in the face of sudden changes, is the first step towards developing lasting mental strength.

Another key aspect of mental strength is stress management. Living under prolonged stress can wear down even the strongest person, making it difficult to think clearly or make rational decisions. For this reason, it is essential to develop techniques that allow you to manage stress effectively. Deep breathing, meditation and exercise are powerful tools for maintaining stress control. But beyond these techniques, it is essential to cultivate a positive mental attitude. Instead of focusing on problems, learn to focus on solutions. The ability to see an opportunity in every challenge, to find the positive side even in the most difficult situations, is what will allow you to overcome obstacles without getting overwhelmed. Mental strength is also closely linked to discipline. In a crisis, discipline isn't just about following rules or procedures, but also about maintaining a routine and following plans you've made in advance.

Discipline helps you stay focused, not deviate from the established path, even when things get difficult. It's what allows you to keep moving forward, step by step, even when the goal seems far away. But discipline is not something that is imposed from the outside; it is a trait that must be cultivated internally, through constant practice and determination. Part of mental strength is also maintaining control of your emotions. Emotions are powerful, and if left unmanaged, they can easily take over and negatively influence your decisions. This doesn't mean suppressing emotions, but rather learning to recognize and manage them so they don't interfere with your ability to function effectively. Emotional awareness is key: Knowing how you feel, understanding the source of these emotions, and finding constructive ways to express them is essential to maintaining mental balance. Finally, developing mental toughness requires a growth mindset. You must be willing to learn from your mistakes, to see difficulties as opportunities to improve and to never give up in the face of obstacles. Mental strength is not only the ability to resist, but also the ability to get back up stronger after every fall. It is the belief that, regardless of what happens, you have the strength within you to overcome any difficulty.

Keeping spirits high in prolonged crises

In a prolonged crisis, when the days seem to blend into one another and uncertainty becomes the new normal, keeping spirits high can be one of the most difficult challenges. Yet, it is precisely in these moments that state of mind and morale become fundamental elements to guarantee the safety of your home and the well-being of those who live there. Keeping calm under pressure, staying motivated and continuing to believe that things can get better is not just a question of mental strength, but a survival strategy that has a direct impact on the ability to face difficulties with clarity and resilience. When the crisis is prolonged, mental and physical fatigue begins to be felt. The days of uncertainty accumulate, and with them the tension and stress. It is normal, in these moments, to feel the weight of the situation and start to doubt your ability to cope. But this is where the importance of keeping spirits high comes into play. Your attitude, your ability to find hope and motivation even in small things, can make the difference between surviving and thriving, between giving in to despair or finding the strength to keep fighting. A key aspect of keeping your spirits high is time management and creating a routine. Even if external circumstances are out of control, having daily structure provides a sense of stability and predictability. Knowing that there are moments dedicated to work, self-care, rest and socialization, even if only virtual, helps keep the mind occupied and prevent the feeling of chaos and disorder.

Routine becomes an anchor to hold on to, a way to maintain discipline and to remind yourself that, despite everything, you can still exercise some control over your life. Human contact, even if limited by circumstances, is another key factor. We are social creatures by nature, and prolonged isolation can have devastating effects on morale. Even under conditions of physical distancing, finding ways to stay connected with others is essential. Conversations, laughter, sharing experiences and feelings help keep the feeling of belonging alive and remember that we are not alone in this struggle. Mutual support, even just a phone call or text, can lift your spirits and provide the comfort you need to face another day. Hope is an element that must never be underestimated. Even in the darkest moments, maintaining a positive vision of the future is essential. This doesn't mean denying reality or ignoring difficulties, but rather cultivating the belief that, eventually, things will get better. Hope is a driving force that pushes you to keep fighting, not to give up, to find solutions even when everything seems to be going wrong. It's what allows you to see beyond the immediate horizon and work towards a better future, even when the present is difficult. Finally, it is important to take care of yourself, not only physically but also mentally.

Find moments of calm, even brief ones, to recharge your energy. Meditation, reading, listening to music or simply contemplating nature can help reduce stress and bring the mind back to a state of balance. Self-care is an act of resistance, a way to preserve your ability to react and face challenges with clarity and strength. In conclusion, keeping spirits high during a prolonged crisis is not just a matter of mental toughness, but an active strategy to ensure the safety and well-being of your home. With the right attitude, social support and self-care, you can face difficulties with the belief that, even if the road is long, you have the strength and determination to overcome any obstacle. Your ability to keep your spirits high is one of the most powerful weapons at your disposal, a strength that allows you to turn crisis into an opportunity to grow, to strengthen bonds, and to come out the other side stronger than before.

CHAPTER 11:

Monitoring and security systems

Imagine always being one step ahead, ready for any eventuality, with the knowledge that every corner of your home is under control. Technology today offers us incredible tools to keep our homes safe, but installing a few cameras or an alarm is not enough to sleep soundly. The key is to integrate these tools with a well-planned strategy and in-depth knowledge of the threats and vulnerabilities you may face.

The first line of defense is awareness. Knowing what's happening inside and outside your home is essential. Modern monitoring systems, such as high-definition surveillance cameras and motion sensors, allow you to have eyes everywhere, even when you are not physically present. But it's not enough to install them; they must be strategically placed to cover all critical access zones and blind spots. It is important that you are able to access these systems in real time, from anywhere, via your smartphone or computer, to react immediately to any anomaly.

A good security system must also have effective alarm systems, linked directly to law enforcement or a private security company. However, the alarm doesn't just have to sound when someone forces a window or door. Today, there are advanced solutions that can detect suspicious movements even outside your home, alerting you before an intruder has the chance to enter. This gives you the advantage of time, a crucial element in any defense operation.

But it's not enough to rely on technology alone. You need to create a protocol for every possible scenario. What to do if the alarm system goes off while you are at home? How to react if someone tries to enter at night? Every family member should know exactly what to do in an emergency. Doors must be solid, preferably reinforced, with high-quality locks, and windows must be protected by burglar-proof glass. If you have a garden or outdoor space, consider installing high, sturdy fences, perhaps with integrated sensors that detect any attempt to climb over. Another crucial aspect is access control. Don't allow anyone to enter your home without careful consideration. Video intercoms, electronic locks and biometric recognition systems can be very effective tools to ensure that only authorized people can access. And remember, even the best security system is useless if it is not used correctly. It's easy to forget to set the alarm or leave a window open; these small errors can compromise the safety of the entire home.

Don't forget, then, the importance of good lighting. Burglars hate to be seen, so make sure the exterior of your home is well lit, especially in the most vulnerable areas. Motion-activated lights can be a great deterrent, making anyone with bad intentions think twice before approaching. Making your home a safe place requires dedication, attention to detail and a proactive approach. It's not just about protecting material things, but about creating an environment where you and your family can feel safe and secure, every day. With good planning, smart use of technology, and a vigilant mindset, you can turn your home into a fortress, ready to withstand any threat.

Set up and keep an eye on home security cameras

Imagine always having full control of the situation, knowing exactly what is happening around you, leaving nothing to chance. This is the mindset you need to adopt when it comes to protecting your home. Home security cameras are not just technological tools, but true extensions of your senses, capable of giving you the situational awareness that is crucial to ensuring the safety of your living space. When it comes to installing and monitoring these systems, every detail counts, because the margin for error must be reduced to zero. It's not just about placing a camera and hoping it does its job; it's about thinking ahead, analyzing the critical points in your home, and making the most of every tool at your disposal.

The first thing to consider is coverage. Every corner of your home, especially the main entrances and ground floor windows, needs to be monitored. You must not leave any blind spots, any areas unsupervised, because that is precisely where the danger could come from. Cameras need to be placed at strategic heights, out of reach of anyone, but low enough to capture precise details. Also, don't forget the importance of lighting: cameras with night vision are essential, but the right ambient light can also make a difference, making it more difficult for anyone to move around unnoticed.

Once installed, it's not enough to forget about the cameras and let them do everything on their own. Monitoring is key. You must be able to access the images at any time, from anywhere. Today, technology allows you to do it directly from your smartphone, so you can control your home even when you are away. And it's not just about watching recorded videos; you need to be able to receive real-time notifications if suspicious movement is detected. This gives you the advantage of reaction speed, which is crucial in any security operation.

It's also important to test your cameras regularly. Don't wait for something to happen to realize that a camera isn't working, or that the camera angle isn't right. Perform periodic checks, make sure everything is in order, and make adjustments if necessary. Remember, prevention is your best weapon. And if you really want to make sure everything works as expected, simulate emergency situations. Do some tests, put yourself in the shoes of someone who might try to breach your security, and try to understand where you could improve. Another crucial element is video storage. Don't underestimate the importance of having records that you can consult if needed. Opt for systems that allow you to store videos securely, perhaps in the cloud, so as not to lose them in the event of hardware damage. And make sure that only you, or people you trust, can access these recordings. Data security is just as important as physical security. Remember that technology evolves and so should your security systems. Keep up to date with new solutions available on the market, because what is cutting-edge today could be outdated tomorrow. Never settle for a system that works well, always try to improve it, update it, make it even more effective. The security of your home is an ongoing commitment, a mission that requires dedication, attention and a proactive approach. Only in this way can you truly feel safe, knowing that you have done everything possible to protect what is dearest to you.

Alarms, motion sensors and smart home integrations

Imagine your home as a perimeter to be defended, an outpost that requires meticulous strategy and constant surveillance. Every detail counts, every decision can make the difference between being vulnerable and having full control of the situation. Alarms and motion sensors are not just accessories, but vital tools in an integrated home defense system, and how you use them determines the effectiveness of your security. A good alarm system is like a tireless sentry, always vigilant, ready to take action at the slightest sign of danger. It should never be considered as a simple deterrent measure, but as the first line of defense. The installation must be carefully thought out, covering all access routes, leaving nothing to chance. Doors and windows must be equipped with sensors that detect any attempted forcing. These sensors don't just trigger an audible alarm, they immediately send notifications to your smartphone, allowing you to react instantly, wherever you are. And it's not just about responding to the alarm, but about acting with the awareness that time is your most precious resource in these situations.

Motion sensors represent another essential pillar. These devices not only constantly monitor their surroundings, but are able to distinguish between harmless movement and a potential threat. It is essential that they are positioned in strategic areas, such as corridors, main entrances and places of forced passage. They must work silently, without false alarms that can reduce your readiness. Modern technology allows you to adjust the sensitivity of these sensors, so as to prevent pets or small movements from triggering an alarm. The goal is to maintain constant vigilance without compromising daily tranquility.

Integration with smart home technologies adds an extra layer of security. Imagine being able to control every corner of your home simply by speaking to a virtual assistant or using an application on your phone. The lights that turn on automatically when movement is detected, the cameras that activate and send images in real time, the smart locks that you can manage remotely: all this is no longer science fiction, but reality. And it's not just about convenience, but about creating an integrated system that responds in unison to any potential threat. Automation allows you to simulate your presence at home even when you are away, a deterrent element that can discourage any intruders before they even decide to act.

But having a sophisticated system isn't enough if you don't know it inside out. You need to understand how it works, be able to verify its efficiency, and above all, know what to do in every possible scenario. Regular maintenance is essential; a security system must always be ready to function at its best. Each component, from the alarm to the sensors, must be tested periodically to ensure that there are no malfunctions or connectivity problems. You must have a clear action plan, know how to react in the event of an alarm and have direct communication with those who can intervene to help you, whether they are the police or a private security company.

Ultimately, making your home a safe place isn't just a matter of technology, but of mindset. You must be proactive, always alert, ready to improve and adapt your defenses based on new threats that may emerge. You can't afford to let your guard down, because security is a responsibility that requires constant commitment. And when everything is in its place, when you know that every corner of your home is protected and monitored, only then can you relax knowing that you have done everything possible to protect what you love most.

Create a neighborhood watch network

A neighborhood watch network is much more than just a group of people keeping an eye on what's going on. It is a truly coordinated operation, where each member of the neighborhood becomes a sentinel, alert and ready to intervene if necessary.

The first step is to build trust. Without trust, no surveillance network will ever function. You need to know your neighbors, build strong relationships, and share a common goal of keeping the neighborhood safe. This requires open and regular communication, regular meetings to discuss concerns and possible threats, and a willingness to work together for the common good. Once this foundation is established, you can begin to develop a surveillance plan. Every house, every street and every corner of the neighborhood must be under control. This does not mean turning the neighborhood into a militarized zone, but rather creating a system where everyone knows what to watch, how to report suspicious activity and how to react in case of an emergency.

Technology plays a crucial role in all this. Today, you can integrate your surveillance network with digital tools that facilitate communication and monitoring. Instant messaging groups, neighborhood safety apps, and even shared cameras covering common areas can be extremely effective tools. But technology is only useful if it is used correctly. Each member of the network must be trained on how to use these tools, what to look for, and how to report information in a clear and timely manner. A surveillance network must not only be reactive, but also proactive. This means that you don't just intervene when something goes wrong, but you constantly work to prevent threats, identifying potential vulnerabilities and addressing them before they can become a problem.

It is important to establish standard operating procedures. When suspicious activity is reported, each member must know exactly what to do: who to call, how to document the event, and what steps to take to protect themselves and others. This requires constant coordination and regular training, so that everyone is always ready to intervene. And it's not just about protecting the neighborhood from strangers. A neighborhood watch network must also be attentive to what is happening within the community, recognizing signs of internal problems, such as situations of abuse or distress, and knowing how to intervene appropriately.

An effective neighborhood watch network creates a sense of community that goes beyond simple safety. When you know your neighbors have your back, when everyone works together towards a common goal, it creates an environment where everyone feels safer and more protected. And this leads to a more cohesive neighborhood, where people know each other, trust each other and are willing to intervene when necessary. Creating a neighborhood watch network is no easy task, but with the right commitment and dedication, you can transform your neighborhood into a safe stronghold, where every member of the community is an integral part of the common defense. It all comes down to being prepared, being aware and leaving nothing to chance. A neighborhood watch network is an extension of your desire to protect what you hold dear, and with the right mindset and the right tools, you can make a difference, not just for your home, but for your entire neighborhood. It's a mission that requires commitment, but in exchange offers peace of mind and the certainty that, together, you can face any threat.

CHAPTER 12:

Sharing resources and community

A truly safe home is one that is part of a strong and supportive community, where sharing resources is not just an act of generosity, but a survival strategy. The ability to connect with others, collaborate and support each other becomes fundamental to face difficulties with a collective strength that far surpasses individual strength. Sharing resources in a community can make the difference between success and failure in managing a crisis. When resources are scarce, a collaborative approach allows you to optimize what is available, reducing waste and ensuring that basic needs are met for all. This approach not only increases individual resilience, but strengthens the social fabric, creating bonds of trust and solidarity that are crucial in times of difficulty. In a community where resources are shared, everyone contributes and benefits, creating a balance that can sustain everyone for longer periods than would be possible alone.

Building a community based on sharing takes time and effort, but the benefits are immense. The first step is to establish a communication network with neighbors and identify available resources. This could include sharing food, water, tools, medicine, or even specific skills and knowledge. Knowing who has what and who is willing to share creates a map of resources that can be vital in an emergency situation. But it's not just about the exchange of material goods; it's also about emotional and moral support. Knowing that you can count on someone, that you have someone to lean on, reduces stress and reinforces the sense of security. The key to effective resource sharing is trust. Trust is built through transparency, open communication and reciprocity. It is important that every member of the community feels respected and that resources are distributed equitably. This not only ensures that everyone's needs are met, but also strengthens the sense of belonging and shared responsibility. A community that supports each other is much better prepared to resist adversity, because it does not rely on the strength of a single individual, but on the collective strength of all its members. Another fundamental aspect of resource sharing is the ability to plan and coordinate.

Resources must not only be shared in response to a crisis, but must be managed proactively. This means working together to create joint stockpiles, establish shared emergency plans and train to respond in a coordinated manner.

This type of preparation not only makes the community more resilient, but also creates a culture of prevention and readiness that can make a difference when crisis hits. Finally, sharing resources within a community is not limited to the moment of emergency, but has a lasting impact. Once the crisis has passed, the bonds created through sharing and collaboration continue to strengthen the community. People who have worked together to overcome difficulties emerge from the crisis with a sense of unity and mutual trust that transforms the concept of a safe home. Your home is no longer just a physical refuge, but becomes part of a larger safety net, where the strength of the community protects and supports you. Ultimately, making your home a safe place in times of crisis means embracing the power of community and sharing resources. Your strength lies not only in the walls that surround you, but in the ability to collaborate, support and be supported by others. In a united community, every home becomes safer, because it is never alone. Sharing resources is an investment in collective resilience, a strategy that transforms vulnerability into strength and uncertainty into security. By sharing resources and building strong bonds, you create a safety net that makes your home, and those of your neighbors, a bulwark against any storm.

Collaborate for mutual help with neighbors

In a world where uncertainty can come in many forms, the safety of your home depends not only on the walls that surround you or the protection systems you have installed. True security extends beyond your front door, rooted in the ability to collaborate effectively with your neighbors. Mutual help between neighbors is not only a principle of good coexistence, but a fundamental strategic element for dealing with any crisis. In difficult situations, the strength of a cohesive community ready to support each other can make the difference between facing a challenge alone or overcoming it together, with greater effectiveness and fewer risks. Collaboration with neighbors begins long before a crisis strikes. Joint preparedness is key to building a support network that can activate quickly in an emergency. This means going beyond simple everyday courtesies and building relationships based on trust, open communication and mutual understanding of each other's needs and available resources. Knowing the strengths, skills and vulnerabilities of those who live next to you allows you to plan more efficiently and respond in a coordinated manner to any unexpected event.

Communication is the first step towards effective collaboration. Establishing clear and reliable communication channels, whether through regular meetings or via messaging groups or two-way radios, ensures everyone is aligned and ready to respond quickly when needed. Sharing information and continuously updating on the neighborhood situation, available resources and possible threats allows everyone to be better prepared and act promptly. In times of crisis, the speed and accuracy of information are crucial, and a well-structured communication network can prevent panic and facilitate informed and coordinated decisions. But collaboration isn't just about managing resources or information; it is also about developing a sense of collective responsibility. In a well-functioning community, everyone has a role to play and a share of responsibility in ensuring the safety and well-being of all. This can mean helping each other with practical tasks, such as sharing equipment or managing neighborhood safety, but it can also translate into moral and psychological support. Knowing that you are not alone, that you can count on someone in case of need, strengthens not only physical security but also emotional resilience, which is equally important in moments of prolonged stress.

During an emergency, collaboration can take many forms. From sharing resources such as food, water or medicines, to creating vigilance groups to monitor neighborhood safety, to the simple act of caring for the elderly or the most vulnerable. These actions, while seeming small, help build a strong and resilient community, able to withstand difficulties and emerge more united after the crisis. Cooperation is not just a practical necessity; it is a way to build stronger bonds, based on trust and mutual respect. However, to collaborate effectively it is essential that there is clear planning and that all community members are aware of the action plans. This may include holding regular meetings to discuss emergency response strategies, creating checklists to prepare in advance, or assigning specific roles in the event of a crisis. A well-organized community is a prepared community, and this preparation can make the difference between an effective response and a chaotic one. In conclusion, making your home a safe place requires not only individual measures, but an active commitment to building strong relationships and a culture of collaboration with your neighbors. Mutual help is not just a strategy to overcome difficulties, but a way to strengthen the social fabric of your neighborhood, creating an environment in which everyone's safety becomes everyone's safety. With a cooperative mindset and an ongoing commitment to nurturing these bonds, you can ensure that your home not only withstands crises, but thrives through the strength and solidarity of the community around it.

Create a community of support

Creating a community of support around your home is one of the most strategic steps you can take to ensure the safety and well-being of you and your family. When it comes to making your home a safe place, you can't just stop at physical security or emergency reserves. True security comes from the knowledge that you are not alone, that you can count on a network of people ready to support each other in times of difficulty. Creating a supportive community is not a simple act of socialization, but an investment in collective resilience, a safety net that extends beyond your property, encompassing the neighborhood and, ideally, the broader community. The first step to building this community is open and continuous communication. It is essential that you know your neighbors and that they know you. This doesn't just mean exchanging warm greetings, but also taking the time to build authentic relationships based on trust and mutual understanding. Finding out who they are, what their skills are, and what resources they can make available in an emergency allows you to identify the community's strengths. An effective support network is based on complementarity of skills and resources, and dialogue is the key to discovering how everyone can contribute to collective well-being.

A supportive community works best when there is a clear sense of shared purpose. It's not just about helping each other with small daily problems, but about being ready to respond in a coordinated and determined way to crises. This requires advance planning. Meeting regularly to discuss potential risks the community could face, whether natural disasters, safety issues or health emergencies, and planning joint strategies to address them, allows everyone to know what to do and who to turn to when the need arises. Knowledge and preparation not only reduce panic in critical situations, but also increase the effectiveness of the collective response.

Collaboration is the driving force of a supportive community. When you create an environment where people feel encouraged to share their ideas, propose solutions and actively participate in community life, a sense of belonging develops that strengthens bonds between members. This sense of belonging is crucial because it motivates people to protect not only themselves, but also their neighbors. Knowing that someone is watching over you, and that you are doing the same for them, builds a level of trust that is critical in any security environment. Creating a community of support does not just mean reacting to emergencies, but also building a culture of resilience in everyday life. It means encouraging a lifestyle that values preparation, mutual help and solidarity. This can be done through organizing community activities that strengthen social cohesion, such as neighborhood events, joint projects or simply regular meetings to discuss issues that affect everyone. These activities not only strengthen bonds between people, but help create an environment where everyone feels involved and responsible for the safety and well-being of the community. Ultimately, a supportive community is stronger the more inclusive it is. Welcoming new members, listening to diverse perspectives, and ensuring everyone has a voice in decision-making strengthens the community and makes it better able to meet challenges.

Diversity of experiences and skills is a resource that enriches the community and increases its ability to resist and adapt to changing circumstances. In conclusion, creating a community of support around your home is not just a safety measure, but a strategic choice that can make a difference in times of crisis. A united community is a strong community, capable of facing any challenge with determination and solidarity.

Your home will be truly safe not when it is isolated from the rest of the world, but when it is part of a support network that values collaboration, preparation and mutual respect. Building this community requires commitment, but the resulting benefits go far beyond simple physical protection, helping to create a safe, peaceful and sustainable living environment for all.

Secure exchange and sharing of resources

When it comes to ensuring the safety of your home, it is essential to approach the problem with the same care and precision that we would apply on a critical mission. Every detail, every decision counts. The key is to take a strategic and proactive approach. First, understand that security is not just a matter of physical protection, but also of managing and securely exchanging resources. Every element of our home, from the front door to the communication systems, must be considered as part of a complex system in which each component must function optimally and safely. To begin, let's look at the importance of a good access and control system. Doors and windows must be equipped with robust locks and, if possible, with warning systems that can signal unauthorized access. Investing in modern security devices, such as surveillance cameras and motion sensors, can provide an extra layer of protection. However, it's not enough to just install them; It is critical to ensure these devices are updated and properly configured. An effective security system must be able to respond in real time to any anomaly.

Security is not just limited to physical assets, but also extends to the protection of personal and digital information. In an age where data can be vulnerable, it is essential to take measures to protect sensitive information. Using strong passwords and changing them regularly, using up-to-date security software, and being careful with internet communications are all crucial practices. Home Wi-Fi networks must be protected by secure, encrypted passwords to prevent unauthorized access. Preparation and planning are equally crucial. Having a clear and well-defined emergency plan for

75

your family can make the difference in a critical situation. This includes knowing how to react in the event of a fire, intruder alert or other emergency. It's helpful to have some sort of "emergency kit" ready and accessible, containing essential items such as first aid kits, important documents and basic survival resources. The concept of exchanging and sharing resources is also of considerable importance. Establishing safe and reliable channels for communicating and sharing resources with neighbors and community members can strengthen overall safety. Collaborating with neighbors and participating in local safety groups can help create a support network that can provide additional assistance and protection if needed. Furthermore, sharing information about threats or suspicious events in the area can improve collective response capacity. In summary, to make your home a safe place, it is essential to adopt a systematic and strategic approach that considers every aspect of security, from physical access to the protection of personal and digital information, up to emergency preparedness and collaboration with community. Each element must work in synergy to ensure effective and continuous protection. The security of our home depends on constant vigilance, meticulous preparation and careful management of resources, just like a well-planned and executed mission.

CHAPTER 13:

Superior Home Defense

In order to afford an ideal home defense, carefully observe every entrance, every vulnerable point. Consider how a possible intruder might try to break into your home and what your weak points are. The key is to think like the enemy, but with a protection-oriented focus. Next, focus on the physical barrier. Doors and windows must be sturdy and well reinforced. A simple padlock is not enough; opt for high security locks that can resist forcing attempts. Also add reinforcements to windows and doors, if necessary, to improve their impact resistance. Furthermore, consider installing an armored door, especially for the main entrances, to further raise the level of protection. Security is not limited to the physical structure of the home. An effective warning system is essential. Invest in a high-quality alarm system and make sure it covers all critical areas of your home, including the most hidden areas. Motion sensors and surveillance cameras can provide an extra layer of security, allowing you to monitor suspicious activity and respond promptly. Make sure the warning system is always functioning and tested regularly.

Another essential component is preparation and planning. Create an emergency plan that involves all members of your family. Make sure everyone knows what to do in the event of an intrusion or emergency. Define safe meeting points and procedures for contacting the competent authorities. Preparation and practice can make the difference between an effective response and a disorganized reaction. Don't forget the importance of communication and collaboration with your neighbors. A good relationship with the surrounding community can be a significant deterrent to bad actors. Attend neighborhood meetings and exchange safety information with other residents. Collective awareness and mutual vigilance increase the protection of the entire area.

Finally, always maintain a vigilant attitude. Securing your home is an ongoing process, not a goal you reach and forget about. Constantly monitor your security measures and adapt your strategies based on new information and changing circumstances. Preparation and prevention are key to keeping your home a safe place, and every step you take toward greater safety is a step toward stronger protection for you and your family.

Strengthen the structural integrity of your home

Every facility has vulnerabilities, and identifying them is crucial to improving overall security. Start by examining the foundation, which is the basis of your protection. Make sure they are solid and free of significant cracks. The foundations must be strengthened if necessary, using appropriate techniques such as strengthening with resin injections or reinforcing with steel bars. A solid foundation not only ensures the stability of the house, but also offers a robust defense against any attempts at forced entry. The exterior walls of your home must be equally sturdy. Check that they are well consolidated and resistant to impacts. A strengthening of the walls can include the use of high-quality materials and modern construction techniques that improve their ability to resist external stresses. Remember that walls serve not only as a physical barrier, but also as protection against intrusion.

Don't neglect the roof. A well-maintained and solid roof is essential to the structural integrity of your home. Make sure there are no leaks or obvious damage. If necessary, reinforce the roof with resistant materials and check its condition regularly. A roof in good condition not only prevents water infiltration, but also contributes to the overall stability of the structure.

Windows and doors are other critical points to consider. Reinforce these areas with high-security locks and additional reinforcements, such as security bars or burglar-resistant glass. Windows must be impact resistant and designed to impede any attempted entry. Investing in sturdy window and door solutions not only improves security, but also increases your peace of mind. Finally, the structural safety of your home cannot ignore the quality of the materials used. Choose resistant and reliable materials for all parts of your home. The durability of the materials contributes significantly to the overall strength of the structure and protection against external factors.

Bulletproof and explosion-proof protection options

Bulletproof and explosion-proof protection options are choices that can provide superior safety. Imagine your home as a fortress that must resist extraordinary situations. The approach must be scrupulous, every detail must be meticulously analyzed and implemented.

Let's start with body armor. These systems are designed to offer an effective barrier against bullets and shrapnel, and are made using advanced materials such as bulletproof glass and Kevlar or carbon fiber armor. Installing bulletproof windows can be a key preventative measure in high-risk settings. These glasses are multi-laminated and reinforced to resist impacts and gunshots. Not only do they offer significant protection, but they also improve the overall security of the home, increasing the level of defense against external threats. For doors, you can opt for armored models, specifically designed to resist break-in attempts and attacks. These doors are built with special materials that absorb and dissipate the energy of impacts, making unauthorized access very difficult. Another option is to install bulletproof armor in the most vulnerable areas of the house, such as main entrances and at-risk windows.

When it comes to explosion protection, which deals with defending against explosions and high-intensity fires, it is important to consider the use of fireproof materials and specialized insulation systems. Walls and ceilings can be reinforced with fire-resistant materials, while exterior cladding and structures can be treated to resist explosions or fires. Fireproof materials such as fire-resistant gypsum board or ceramic coatings can help protect your home from high temperatures and shock waves. Furthermore, it is advisable to implement an early detection and warning system for fires and explosions. Smoke and gas sensors, combined with an alert system that immediately notifies authorities and emergency services, can provide a rapid response in the event of an accident.

Adopting these measures requires a considerable investment, but the security of your home is invaluable. Each element of bulletproof and explosion-proof protection must be carefully integrated, considering specific needs and potential risks. The key is to plan and prepare comprehensively, not only to address immediate threats, but also to ensure ongoing protection over time. With careful preparation and strategic implementation, you can transform your home into a bastion of security, ready to withstand even the most extreme challenges.

Creation of a covered safety area

Creating a covered safety area begins with choosing the ideal location inside your home. This space must be located in a strategic position, preferably in the center of the house, away from windows and doors, to minimize risks from the outside. If possible, select an area that will be easily accessible to all family members in the event of an emergency. This not only ensures that everyone knows where to go, but also that they can do so quickly and without panic. Once the location has been identified, it is essential to harden the area to ensure it can withstand a variety of threats. Start with reinforcing the walls and ceiling. Use strong, long-lasting materials, such as reinforced concrete or steel beams, to create a structure that can withstand impacts and external stresses. If your budget allows, also consider installing bulletproof elements to further increase protection.

The next step is setting up the area. This area must be equipped with all materials and resources necessary to support your loved ones during an emergency. Always keep non-perishable food, water and basic medical supplies available. Make sure there is a reliable source of lighting and that you have a functioning communications system, such as a battery-operated radio, to stay informed of any external developments and to contact emergency services if necessary. Also include a ventilation system to ensure the air within the area remains breathable, especially in the event of emergencies that could affect air quality, such as fires or contamination. The ventilation system must be designed to filter any harmful particles and ensure continuous air flow. Finally, consider the possibility of making the covered safety area also a family meeting point. Establish a clear emergency plan and ensure everyone knows how to get to and use this space. Teach all family members how to access the area and how to use the resources you have made available. Preparation and familiarity with the plan can make all the difference in high-pressure situations. Creating a covered safety area is an investment in peace of mind and protection. It's not just about building a physical shelter, but ensuring that in the event of a major emergency, you and your loved ones have a safe place to retreat and stay safe. With careful planning and precise execution, you can transform a simple space into a bastion of security and peace of mind for your entire family.

CHAPTER 14:

Adaptation to changing circumstances

To ensure your home remains a bastion of safety, it's crucial to approach the issue with the mindset and precision of someone who manages complex and changing situations. Home security is not a static goal, but an ongoing dynamic that requires adaptation and responsiveness. Think of your home as a fortification: it must be able to respond effectively to a variety of threats, which evolve over time. Every element of your security, from locks to surveillance technologies, needs to be regularly evaluated and updated to meet new challenges. There is no single, definitive solution; rather, it is crucial to develop a mindset of constant adaptation and improvement. Start with a thorough vulnerability analysis. Threats change over time, and the techniques used by bad actors continually evolve. Your home security must be designed to be flexible and able to adapt to these variations. This means you need to regularly monitor and update your warning systems and safety technologies. For example, if a new vulnerability is discovered in a type of lock or alert system, it is critical to quickly replace or upgrade those devices.

Monitoring and analyzing external circumstances are equally essential. Neighborhood conditions and community dynamics can change, affecting the level of risk. Being informed about changes in your area, such as increases in crime or the presence of new environmental risks, allows you to adapt your security strategies proactively. Maintain an open dialogue with your neighbors and actively participate in local safety initiatives. This not only helps you stay updated, but also builds a support network that can be invaluable in an emergency. Preparation for unexpected circumstances is another crucial element. Emergencies can occur without warning and require a quick and effective response. Having well-defined and workable contingency plans is essential. This includes knowing how to react in the event of unexpected events, such as fires, floods or intrusions. Create and regularly update an evacuation plan and make sure all family members are aware of the procedures to follow. Another aspect to consider is communications security. In an increasingly connected world, personal information and sensitive data can become targets for attacks. Taking steps to protect your communications and home network is critical. It uses modern technologies, such as encryption and security software, to keep your information safe.

Staying vigilant about new cyber threats allows you to protect your home not only physically, but digitally as well. Finally, the adaptation mentality requires continuous training and updating. Security techniques develop and threats emerge dynamically, and staying informed about home security best practices and innovations is crucial. Investing time and resources to stay current ensures that your security strategy remains effective and relevant over time. In conclusion, to make your home a safe place, it is essential to take an adaptive and proactive approach to security. Your home should be viewed as a dynamic structure that requires regular updates and ongoing vigilance. Only through careful planning, constant monitoring and an open mindset to change can you ensure that your shelter remains a bastion of safety, ready to face any changing circumstances.

Adaptation of the strategy to various situations

Every security strategy must be flexible. Imagine your home as a complex system where each part interacts with the others. The security of doors and windows, for example, must be considered not only in terms of physical robustness, but also of resistance to modern intrusion techniques. This means that a high-quality lock today may not be sufficiently secure tomorrow if lock-picking techniques evolve. Being ready to update security devices and integrate new technologies is crucial to maintaining adequate protection. External circumstances can greatly influence your security strategy. The dynamics of your community, such as the level of crime or changes in social habits, can create new challenges or reduce the risk of certain events.

Monitoring and adapting your measures based on these factors allows you to respond more effectively to threats. For example, if you notice an increase in suspicious activity in your area, it may be time to strengthen surveillance systems or increase collaboration with neighbors. Emergency management is another area where flexibility is essential. Not all emergencies follow a predictable pattern. Whether it is a sudden fire, a flood warning or a cybersecurity crisis, it is essential to have contingency plans that can be quickly adapted to the specific situation. Preparing your family to respond effectively and safely to various types of emergencies greatly increases the chances of keeping the situation under control and minimizing damage. Furthermore, security is not just about immediate protection, but also about the ability to anticipate and prevent future situations. Personal information and digital assets should be protected with the same care you give to the physical security of your home. Cyber threats evolve rapidly and taking appropriate cybersecurity measures is essential.

This includes regularly updating security software, using strong passwords, and handling sensitive information carefully. Finally, the adaptation mentality also implies continuous training and updating. Don't stay static; The security landscape is changing, and to stay one step ahead of threats, it's important to regularly educate yourself on best practices and new technologies available. This will allow you to continually optimize your security strategy, ensuring it remains effective and relevant. In summary, keeping your home safe requires a strategic approach that adapts to changing circumstances and threats. An effective security system is one that not only responds to immediate challenges, but is also ready to evolve and improve over time. Being proactive, flexible and informed allows you to create a protected environment, capable of facing any situation with determination and preparation.

Reaction to new dangers and obstacles

Every home is a microcosm of safety that must respond to a variety of potentially dangerous situations. A key part of this process is the ability to recognize and respond to new dangers. Whether it's a new form of intrusion, a change in the dynamics of your area, or security technology becoming obsolete, your reaction must be quick and effective. When a new threat emerges, the first step is a thorough assessment. This means gathering detailed information and understanding the context in which the threat occurs. For example, if there is an increase in thefts in your area, it is essential to identify common patterns and understand the techniques used by bad actors. This knowledge will allow you to adapt and strengthen existing security measures. It is not enough to install a new warning system; you also need to ensure that it is integrated effectively with the other components of your security system and that all family members are informed on how to best use it. Furthermore, reacting to new obstacles also involves constantly reviewing security resources and technologies. For example, intrusion techniques develop over time, and what was considered safe yesterday may no longer be safe today. It is important to maintain an open mind and constant vigilance regarding innovations and new threats. Updating security systems, such as surveillance cameras and electronic locks, and ensuring they are equipped with the latest protection technologies is a crucial step in thwarting new types of attacks. Another essential aspect in reacting to new dangers is preparation and planning. This means not only having contingency plans for known threats, but also developing a resilience mindset that allows you to face the unexpected. Creating emergency scenarios and simulations that include new threats or obstacles can help prepare your family to respond quickly and in a coordinated manner. These exercises can be vital in ensuring that, in the event of a real emergency, everyone knows exactly what to do and how to react. We must not forget the importance of communication and collaboration with the community.

Sometimes, new dangers and obstacles can be better addressed through a collective approach. Collaborating with neighbors and participating in local safety groups allows you to exchange information and strategies, and get additional support in case of emergencies. A well-organized support network can offer resources and assistance that are beyond the capabilities of any individual security system. Finally, it's important to remember that security is a continuous and evolving process. The ability to adapt to new dangers and obstacles requires constant vigilance and a commitment to continuous improvement. Investing time and resources into regularly evaluating and updating your security measures not only strengthens the protection of your home, but also ensures that you are prepared to face any future challenges. In summary, to keep your home safe in the face of new dangers and obstacles, it is essential to take a proactive and adaptive approach. Being informed, prepared and ready to respond to new threats will allow you to keep your haven safe and secure, ensuring it remains a bastion of security in every situation.

Gain insights from real bug-in incidents

When it comes to securing your home, it's crucial to understand that practical experience and knowledge from real incidents can provide valuable insights into strengthening your protective measures. Managing a critical event, known as a "bug-in," is a field where every detail counts, and learning from the experiences of those who have faced similar situations can make the difference between an effective response and an inadequate strategy. A bug-in is an episode in which you find yourself forced to remain inside your home, either due to unforeseen circumstances such as natural disasters, security crises or pandemics, or other emergencies. In these cases, home is not just a place of refuge, but becomes the center of your survival strategy. Learning from the experiences of those who have handled similar situations offers practical insight into how to prepare and respond appropriately. Every real bug-in incident offers key lessons.

Observing and analyzing how people have addressed issues such as resource management, communications security and the effectiveness of warning systems can provide crucial insights into how to improve your preparedness. For example, in situations where resources such as food and water become limited, advance planning and inventory management are essential. Past experiences show that having a reserve of essential goods, not only for the short term but also for an extended period, can make a difference. Additionally, during a bug-in, the home's internal security takes on even greater importance. Real incidents often show that the most sophisticated security systems are those that not only prevent intrusions, but also enable a rapid and coordinated response in the event of an emergency. Evaluating the strategies adopted by those who have faced such situations helps you understand which systems work best and how you could adapt your security devices to the specific needs of your home. Communications is another crucial aspect that emerged from real bug-in events. The ability to maintain effective communication with family members and, if possible, the outside community, is vital.

In many emergency situations, communications lines can become congested or unavailable, so it is critical to have alternative plans and backup communications tools. Past experiences have shown that clear, well-planned communications are critical to coordinating responses and managing emergencies effectively. Finally, continuously monitoring and updating your security measures, based on knowledge gained from real incidents, is essential. Home security is not static and must be constantly adapted based on new information and evolving threats. Past events offer useful insight into how to improve preparedness, how to identify vulnerabilities, and how to optimize your response to critical situations. In summary, to make your home a safe place, it is crucial to draw on knowledge gleaned from real bug-in incidents. These experiences offer valuable lessons on how to manage resources, improve safety and maintain communications in emergency situations. Integrating this information into your preparation will allow you to face any situation with greater confidence and preparation, transforming your home into a true fortress of protection and resilience.

CONCLUSION

Throughout this book, we have explored the concepts of home security and emergency preparedness in depth, approaching these issues with the same seriousness and precision that characterize the training of Navy SEALs. This journey has led us to understand that the safety of our home and the ability to survive in critical situations are not the result of a single intervention, but rather of a continuous and multidimensional process. We began by addressing the importance of "bugging in," a fundamental concept that, if well understood and applied, can make the difference between success and failure in crisis situations. Staying in your home during an emergency, rather than fleeing into the unknown, offers significant benefits. Familiarity with the environment, control of resources and the ability to organize effective defenses are all elements that contribute to making your home a safe fortress, capable of protecting you and your loved ones from external threats.

We then explored techniques for conserving essential resources, such as food and water, and the importance of having access to reliable, self-sufficient energy. In an increasingly unpredictable world, the ability to autonomously manage these resources becomes crucial. Preparation is not limited to stockpiling, but also requires careful management and constant maintenance. Rotating food supplies, safely storing water, and installing independent energy systems, such as solar panels and generators, are practices that need to be integrated into your daily routine to ensure long-lasting resilience.

Additionally, we analyzed the home's physical security measures, from doors and windows to surveillance systems and exterior lighting. Home defense is a set of strategies that must work in synergy, combining technology, prevention and reactivity. It's not just about installing alarms or cameras, but about creating an environment where every detail is designed to deter and neutralize threats, while at the same time maintaining a sense of normality and serenity in daily life. But, preparation is not just a matter of tools and technologies; it is also, and above all, a question of mentality. The Navy SEALs teach us that mental resilience is as important as physical preparation. In a crisis situation, staying calm, making clear decisions and adapting to changing circumstances are vital skills that can make the difference between survival and failure. Your home, however well protected, is only part of the equation: the real key element is you and your ability to face challenges with determination and readiness.

This book is not just a survival manual, but a guide to living in a more conscious and prepared way. Home security should not be seen as a burden or paranoia, but as an investment in your quality of life. Being prepared means you can live with the peace of mind of knowing that, whatever happens, you have done everything you can to protect yourself and your loved ones. Remember that security is a dynamic and ever-evolving process. Threats change, technologies update and our needs transform. It is essential to remain vigilant, constantly update our strategies and adapt to new scenarios. Only in this way can we ensure that our home remains a safe haven, capable of withstanding any challenges the future may bring.

Made in the USA
Columbia, SC
20 September 2024

42643546R00052